第六届亚洲政党国际会议开幕式
At the opening ceremony of the 6th General Assembly of International Conference of Asian Political Parties

首届世界生态安全大会开幕式
At the opening ceremony of the First World Eco-Safety Assembly

束埔寨王国政府总理洪森
Samdech Hun Sen, Prime Minister of the Government of the
Kingdom of Cambodia

马来西亚政府总理纳吉布•敦•拉扎克
Dato' Sri Mohamad Najib bin Tun Haji Abdul Razak, Prime
Minister of Malaysia

菲律宾众议院前议长、亚洲政党国际会议创始主席
何塞•德贝内西亚
Jose de Venecia, Jr., Former Speaker of the House of Representatives of the Philippines, and Founding Chairman of
the ICAPP

尼泊尔民主共和国总理马德哈夫•库玛•尼帕尔
Madhav Kumar Nepal, Prime Minister of Nepal

束埔寨王国政府副总理索安
Sok An, Deputy Prime Minister of the Government of the
Kingdom of Cambodia

中国共产党中央政治局委员、中央书记处书记、中央组
织部部长李源潮
Li Yuanchao, Member of the Political Bureau of the Central
Committee of the Communist Party of China, Member of the
Secretariat of the Central Committee and Minister in charge of
the Organization Department of the CCCPC

亚洲政党国际会议秘书长郑义溶
Chung Eui-yong, Secretary-General of ICAPP

亚太中间党派民主国际(CAPDI)秘书长穆萨•侯赛因•赛义德
Mushahid Hussain Sayed,Secretary-General of CAPDI

菲律宾前总统菲德尔•瓦尔德斯•拉莫斯
Fidel Valdez Ramos, Former President of the Philippines

印度尼西亚前总统梅加瓦蒂•苏加诺
Mrs. Megawati Sukarnoputri, Former President of Indonesia

塞舌尔共和国外长让•保罗•亚当宣读詹姆斯•米歇尔总统贺信并致辞
Congratulatory letter from James Michel, President of Seychelles, to be read by Jean–Paul Adam, Foreign Minister of the Republic of Seychelles

巴基斯坦总统特使穆罕默德•阿扎姆•汗•斯瓦蒂
Mohammad Azam Khan Swati, Special Envoy of the President of Pakistan

拉美加勒比政党常设代表大会副主席古斯塔沃•莫雷诺
Gustavo Moreno, Vice President of Permanent Conference of Latin American and Caribbean Political Parties

菲律宾众议长菲利西亚诺•贝尔蒙特
Feliciano Belmonte, Jr., Speaker of the Philippine House of Representatives

国际生态安全合作组织总干事蒋明君
Jiang Mingjun, Director-General of International Eco-Safety Cooperative Organization

联合国副秘书长、坦桑尼亚国土住房部部长、国际生态安全合作组织联合主席安娜·蒂贝琼卡博士
Anna Tibaijuka, Under-Secretary-General of the UN, Minister of Lands, Housing and Human Settlements Developments of Tanzania, and Co-Chairman of IESCO

上海合作组织首任秘书长、国际生态安全合作组织联合主席张德广
Zhang Deguang, First Secretary-General of Shanghai Cooperation Organization, and Co-Chairman of IESCO

蒙古国前总理、国际生态安全合作组织联合主席门德赛汗·恩赫赛汗
Mendsaikhan Enkhsaikhan, Former Prime Minister of Mongolia, and Co-Chairman of IESCO

乌干达共和国总统特使瓦吉多索宣读约韦里·卡古塔·穆塞韦尼总统贺信
Congratulatory letter from Yoweri Kaguta Museveni, President of Uganda, to be read by Charles M. Wagidoso, Special Envoy of the President of Uganda

澳大利亚自由党前主席、副总理克里斯·麦克迪文
Christine Ann McDiven, Former President of Liberal Party of Australia and Vice Premier

马尔代夫共和国总统特使艾哈迈德·拉蒂夫宣读穆罕默德·纳希德总统贺信
Congratulatory letter from Mohamed Nasheed, President of Maldives, to be read by Ahmed Latheef, Special Envoy of the President of Maldives

联合国大学教务长、国际生态安全合作组织常委威瑟琳·波波夫斯基
Vesselin Popovski, Senior Academic Program Officer of United Nations University and Member of the Standing Committee of IESCO

国际生态安全合作组织专家委员会主席、中国建设部前副部长杨慎
Yang Shen, Chairman of Experts Committee of IESCO and Former Vice Minister of the Ministry of Construction

柬埔寨王国政府国务部长郭平
Kol Pheng, Senior Minister of the Government of the Kingdom of Cambodia

柬埔寨王国政府环境部长莫马烈
Mok Mareth, Minister for the Environment of the Kingdom of Cambodia

欧洲PA国际基金会秘书长里约·普拉宁
Rio D. Praaning Prawira Adiningrat, Secretary-General of PA (Europe) International Foundation

联合国前副秘书长、国际生态安全合作组织联合主席安瓦鲁尔·乔德哈瑞
Anwarul Chowdhury, Former UN Under-Secretary-General and Co-Chairman of IESCO

中国国际问题研究基金会副理事长董津义大使
Ambassador Dong Jinyi, Deputy Board Director-General of China Foundation for International Studies

国际生态生命安全科学院院长鲁萨克·奥列格
Rusak O. N., President of International Academy of Ecology and Life Protection Sciences (IAELPS)

美国密苏里州堪萨斯市首席环境官丹尼斯·墨菲
Dennis Murphey, Chief Environmental Officer, Kansas City, Missouri, USA

国际生态生命安全科学院副院长斯坦尼斯拉夫·阿巴隆斯基
Stanislav Appollonskiy, Vice President of IAELPS

中国四川省南充市纪委书记胡文龙
Hu Wenlong, Secretary of the Commission for Discipline Inspectionce of Nanchong City, Sichuan Province, China

中国东莞市塘厦镇党委书记、人大主席叶锦河
Ye Jinhe, Secretary of the Tangxia Town Party Committee, Dongguan City, China

首届世界生态安全大会闭幕式
At the Closing Ceremony of the 1st World Eco-Safety Assembly

从左到右：国际生态安全合作组织总干事蒋明君与尼泊尔民主
共和国总理马德哈夫·库玛·尼帕尔，柬埔寨人民党常委、副总
理索安，中共中央政治局委员、中央书记处书记、中央组织部
部长李源潮，柬埔寨副总理、国际生态安全合作组织执行主席
盖博·拉斯米会见后合影
From left to right: Jiang Mingjun, Director-General of International Eco-Safety Cooperative Organization, Madhav Kumar Nepal, Prime Minister of Nepal, Sok An, Member of Standing Committee of the Cambodian People's Party and Deputy Prime Minister of the Kingdom of Cambodia, Li Yuanchao, Member of the Political Bureau of the Central Committee of the Communist Party of China, Member of the Secretariat of the Central Committee and Minister in charge of the Organization Department of the CCCPC and Keo Puth Reasmey, Deputy Prime Minister of Cambodia and Executive Chairman of IESCO

国际生态安全合作组织秘书长郑庆森宣读
《世界生态安全大会（吴哥）议定书》
Tee Ching Seng, Secretary-General of International Eco-Safety Cooperative Organization, reads out *(Angkor) Protocol of World Eco-Safety Assembly*

安瓦鲁尔·乔德哈瑞大使夫妇祝贺蒋明君总
干事荣获亚洲政党国际会议授予的"为人类
服务特别贡献奖"
Mr. and Mrs. Ambassador Anwarul Chowdhury congratulate Director-General Jiang Mingjun on "Service to Humanity Award" confered by the ICAPP

第六届亚洲政党国际会议主席团授予蒋明君总干事、郑庆森秘书长"为人类服务特别贡献奖",以表彰他们在促进和平发展、维护生态安全方面作出的杰出贡献
Presidium of the Sixth General Assembly of the ICAPP awards "Service to Humanity Award" to Director-General Jiang Mingjun and Secretary-General Tee Ching Seng of IESCO in recognition of their outstanding contribution to promoting peaceful development and maintaining ecological safety

国际生态安全合作组织作为亚洲政党国际会议观察员,在柏威夏见证《金边和平协议》的签署
As International Observer of the ICAPP, IESCO witnesses the signing of *Phnom Penh Peace Agreement* in Preah Vihear

国际生态安全合作组织总干事蒋明君在亚太中间党派民主国际(CAPDI)会议上当选为气候变化委员会首席执行官
Jiang Mingjun, Director-General of International Eco-Safety Cooperative Organization, is elected the CEO of Climate Change Committee of CAPDI at the CAPDI conference

在首届世界生态安全大会颁奖晚会上，塞舌尔共和国总统詹姆斯·米歇尔，联合国副秘书长安娜·蒂贝琼卡博士获 "维护生态安全特别贡献奖"。图为蒙古共和国前总理门德赛汗·恩赫赛汗向安娜·蒂贝琼卡颁奖

"Special Contribution Award for Maintaining Eco-Safety" is conferred to James Michel, President of the Republic of Seychelles, and Anna Tibaijuka, Former UN Under-Secretary-General at the Award-Giving Evening Party of the First World Eco-Safety Assembly. Picture: Mendsaikhan Enkhsaikhan, Former Prime Minister of Mongolia, confers "Special Contribution Award for Maintaining Eco-Safety" on Anna Tibaijuka

在首届世界生态安全大会颁奖晚会上，马尔代夫共和国政府、柬埔寨王国政府、国际生态生命安全科学院获"维护生态安全特别贡献奖"

"Special Contribution Award for Maintaining Eco-Safety" was conferred to the Government of the Kingdom of Cambodia, the government of the Republic of Maldives, and International Academy of Ecology and Life Protection Sciences at the Award-Giving Evening Party of the First World Eco-Safety Assembly

国际生态安全合作组织丛书
International Eco-Safety Cooperative Organization Series

生态安全：
一个迫在眉睫的时代主题
Eco-Safety,
an Extremely Urgent Theme of the Times

（首届世界生态安全大会文集）
(Collected Works of the First World Eco-Safety Assembly)

■ 主编　蒋明君
Editor-in-Chief　Jiang Mingjun

世界知识出版社

图书在版编目（CIP）数据

生态安全：一个迫在眉睫的时代主题：首届世界生态
安全大会文集：汉英对照 / 蒋明君主编． — 北京：
世界知识出版社，2011.4
（国际生态安全合作组织丛书）
ISBN 978-7-5012-4053-1

Ⅰ．①生…　Ⅱ．①蒋…　Ⅲ．　①生态安全-文集-汉、
英　Ⅳ．①X171-53

中国版本图书馆CIP数据核字（2011）第043301号

责任编辑	侯奕萌
文字编辑	王振兴
责任出版	刘　喆
责任校对	陈可望
封面设计	马　莉
书　　名	**生态安全：一个迫在眉睫的时代主题** Shengtai Anquan: Yige Pozaimeijie De Shidai Zhuti （首届世界生态安全大会文集） (Shoujie Shijie Shengtai Anquan Dahui Wenji)
作　　者	蒋明君／主编
出版发行	世界知识出版社
地址邮编	北京市东城区干面胡同51号（100010）
网　　址	www.wap1934.com
电　　话	010-65265923（发行）　010-65233645（书店）
印　　刷	北京康利胶印厂
经　　销	新华书店
开本印张	787×1092毫米　1/16　$24\frac{3}{4}$印张　4插页
字　　数	467千字
版次印次	2011年4月第一版　2011年4月第一次印刷
标准书号	ISBN 978-7-5012-4053-1
定　　价	45.00元

前　言

2010年12月2—5日，亚洲政党国际会议（ICAPP）、柬埔寨王国政府、国际生态安全合作组织（IESCO）、联合国大学、中国国际问题研究基金会在柬埔寨金边市共同主办了第六届亚洲政党国际会议与首届世界生态安全大会。来自90多个国家的120多个政党、议会、政府和国际组织代表团共1000余人出席了会议。大会围绕"和平发展与生态安全"这一主题展开了深入讨论，并通过了《亚洲政党国际会议（金边）宣言》和《世界生态安全大会（吴哥）议定书》。中国共产党中央政治局委员、中央书处记书记、中央组织部部长李源潮率中国共产党代表团出席会议，并作了重要讲话。

第六届亚洲政党国际会议与首届世界生态安全大会为期四天，分开幕式、专题论坛、闭幕式三大部分。在大会开幕之际，联合国秘书长潘基文作了视频讲话；马尔代夫共和国总统穆罕默德·纳希德，塞舌尔共和国总统詹姆斯·米歇尔，乌干达共和国总统韦里·卡古塔·穆塞韦尼，联合国大学副校长武内和彦为大会发来了贺信。柬埔寨总理洪森，马来西亚总理纳吉布·敦·拉扎克，柬埔寨副总理、国际生态安全合作组织执行主席盖博·拉斯米，尼泊尔民主共和国总理马德哈夫·库玛·尼帕尔，联合国副秘书长安娜·蒂贝琼卡博士，联合国前副秘书长安瓦鲁·乔德哈瑞，菲律宾众议院前议长、亚洲政党国际会议创始主席何塞·德贝内西亚，蒙古共和国前总理门德赛汗·恩赫赛汗，上海合作组织首任秘书长、国际生态安全合作组织联合主席张德广等国际政要发表了致辞。

在主旨演讲和专题论坛中，来自世界各国的政党领袖，议会领导人，政府首脑、部长，环境和生态专家以及国际组织、非政府组织的负责人，围绕"气候变化与生态安全"、"生态安全与灾害预警"、"生态安全与国际多边合作"三大主题展开了讨论。

蒋明君总干事在主旨演讲中回顾了国际生态安全合作组织从2006年6月创建以来的第一个五年计划中在预防自然灾害，解决生态危机和灾后重建方面取得的积极成果。他在谈及第二个五年计划时说，国际生态安全合作组织将以第六届亚洲政党国际会议和首届世界生态安全大会为契机，依据《亚洲政党国际会议（金边）宣言》和《世界生态安全大会（吴哥）议定书》中的原则来制定和发展第二个五年计划。他还强调世界各国应求同存异，加强国际多边合作，共同应对气候变化及人为灾害和冲突。

12月4日晚，在世界生态安全大会的闭幕式上，国际生态安全合作组织总干事蒋明君发表演讲称，这次大会通过的《世界生态安全大会（吴哥）议定书》对指导和促进各国的生态建设，实现经济、社会与生态协调发展具有重要意义，将为实现联合国千年发展目标作出新的贡献。

首届世界生态安全大会呈现三大显著特点。一是亚洲政党国际会议成为本届大会主办单位之一。和平发展与生态安全成为第六届亚洲政党国际会议的主要议题。会议闭幕时通过的《亚洲政党国际会议（金边）宣言》，首次将气候变化与生态安全纳入亚洲政党国际会议的主要议题。二是柬埔寨王国政府参与主办本届大会，并有90多个国家的政党、议会、政府和国际组织代表团出席会议，有利于各国将生态安全纳入国家发展战略。三是本届大会发布的《世界生态安全大会（吴哥）议定书》，表达了与会各政党和政府致力于实现生态安全的决心，共同应对生态危机，实现联合国千年发展目标。另外，本届大会还发起创建国际应急救援机构和世界生态安全基金，期望通过各种有效途径提高政府与社会各界的生态安全意识和自然灾害与生态灾难的紧急救援救助能力。

近年来，全球性重大生态灾害接踵而至，如地震、海啸、飓风、火山喷发；洪水、干旱、泥石流、森林大火、土地荒漠化、沙尘暴、水资源污染、环境污染、工业污染、核污染、湿地锐减、物种灭绝；原油泄漏导致的海洋生态系统破坏；再如艾滋病、非典、禽流感、疯牛病、口蹄疫、霍乱等流行性传染疾病。这些由全球气候变化及人类经济活动引发的生态灾害已对人类生存与国家安全构成严重威胁，且发生的频率和强度仍在不断上升。维护生态安全，消除贫困，应对突发性生态灾害，实现经济、社会、生态的协调发展，需要世界各国政党、议会和政府机构的紧密合作，在多边主义的框架中共同应对挑战。

Foreword

On December 2-5, 2010, the curtain of the 6th General Assembly of International Conference of Asian Political Parties and the 1st World Eco-Safety Assembly that are co-hosted by International Eco-Safety Cooperative Organization (IESCO), International Conferences of Asian Political Parties (ICAPP), Royal Government of the Kingdom of Cambodia, United Nations University and China Foundation for International Studies(CFIS) rose. Political parties, parliaments and governmental delegations from over 90 countries attended the Assembly. After in-depth discussion and excellent dialogue centering on "Ecological Safety & Peaceful Development", *International Conference of Asian Political Parties Phnom Penh Declaration (Angkor) Protocol of World Eco-Safety Assembly* were generated among delegations from various countries. Leading the delegation of Chinese Communist Party, Li Yuanchao, Member of Political Bureau of the CPC Central Committee, Member of the Secretariat of the Cental Committee and Minister in charge of the Organization Department of the CPC Central Committee, attended the Assembly and delivered an important speech.

The 6th General Assembly of (ICAPP) and the 1st World Eco-Safety Assembly lasted four days, including the Opening Ceremony, Thematic Forum and the Closing Ceremony. On the occasion of the Opening Ceremony, Ban Ki-Moon, Secretary-General of the United Nations, delivered a video speech. Mohamed Nasheed, President of the Republic of Maldives, James A. Michel, President of the Republic of Seychelles, Yoweri Kaguta Museveni, President of the Republic of Uganda, and Kazuhiko Takeuchi, Vice-Rector of United Nations University sent their congratulatory letters to the Assembly. At the Opening Ceremony, international political figures such as H.E. Hun Sen, Prime Minister of Cambodia, H.E. Dato' Sri Mohamad Najib bin Tun Haji Abdul Razak, Prime Minister of Malaysia, H.E. Keo Puth Reasmey, Deputy Prime Minister of Cambodia and Executive Chairman of IESCO, H.E. Madhav Kumar Nepal, Prime Minister of Federal Democratic Republic of Nepal, Dr. Anna Tibaijuka, Former UN Under-Secretary-General, H.E. Anwarul K. Chowdhury, Former UN Under-Secretary-General, H.E. Jose de Venecia, Former Speaker of the House of Representatives of the Philippines and Founding Chairman of the ICAPP, H.E. Mendsaikhan Enkhsaikhan, Former Prime Minister of Mongolia, H.E. Zhang Deguang, the First Secretary-General of Shanghai Cooperation Organization, Co-Chairman of International Eco-Safety Cooperative Organization, delivered their speeches.

In the keynote speech and thematic forum, leaders of political parties from various countries in the world, leaders of parliaments, heads of states, ministers, environmental and ecological experts as well as responsible persons of international organizations and non-governmental organizations carried out discussion centering on the three major themes of "climate change and ecological safety", "ecological safety and disaster pre-warning", and "ecological safety and international multilateral cooperation".

In the keynote speech session, Jiang Mingjun, Director-General of IESCO, reviewed the positive achievements in natural disaster and eco-crisis prevention and solution, and post-disaster reconstruction during the 1st five-year plan since its establishment in June 2006. When he talked about the 2nd five-year plan, he said that IESCO would take the 6th General Assembly of ICAPP and the 1st WESA as a good opportunity, follow the principle included in *Phnom Penh Declaration* of ICAPP and *(Angkor) Protocol of World Eco-Safety Assembly*, and develop and make the 2nd five-year plan. He also emphasized that all countries in the world should seek common ground and reserve differences, strengthen international multilateral cooperation, and work together to cope with climate change and other manmade disasters and conflicts.

In the Closing Ceremony of WESA in the evening of Dec. 4th, Jiang Mingjun, Director-General of IESCO, made a speech and stated that *(Angkor) Protocol of World Eco-Safety Assembly*, inspected and approved by the Assembly, would certainly make profound significance on guiding and promoting the ecological construction, and realizing harmonious development of economy, society and ecology for all the countries, and would make new contributions in realizing the UN Millennium Development Goals.

It is reported that the First World Eco-Safety Assembly had three prominent features: first, ICAPP became one of the sponsors of the Assembly. "Peaceful development and ecological safety" became an important topic for discussion of the 6th General Assembly of the ICAPP. The *Phnom Penh Declaration*, which was passed when the 6th General Assembly concluded, brought climate change and eco-safety into the main subjects of the ICAPP for the first time. Second, Government of the Kingdom of Cambodia participated in this Assembly, and delegations of political parties, parliaments, governments and international organizations from over 90 countries were present at the Assembly, which created favorable conditions for eco-safety being included in various countries' development strategies. Third, the *(Angkor) Protocol of World Eco-Safety Assembly*, which was released in this Assembly, represents the common wish of political parties and governments that attended the Assembly to commit themselves to ecological safety, coping with ecological crises and realizing the UN Millennium Development Goals. Moreover, this Assembly also initiated international rescue institutions and World Eco-Safety Fund, aiming to raise the governments, society and the public's awareness on ecological safety and build up the emergency rescue capability to deal with natural disasters and ecocatastrophes by various effective means.

In recent years, we have witnessed the world's worst ecological disasters one after another, such as earthquake, tsunami, hurricane, volcanic eruption, flood, drought, mud-slide, forest fires, land desertification, sandstorm, water resources contamination, environmental contamination, industrial contamination, nuclear pollution, sharp decline of wetlands, species extinction, destruction of marine ecosystem caused by crude oil spill, and epidemic diseases such as AIDS, SARS, bird flu, mad cow disease, foot and mouth disease, cholera. These ecological disasters caused by climate change and human economic activities have posed a serious threat to human survival and national security. And their frequency and intensity are still increasing. Close cooperation of political parties of various countries in the world, parliaments and governmental institutions are needed to maintain ecological safety, reduce poverty, cope with unexpected ecological disasters, and realize coordinated development among economy, society and ecology as well as confronting the challenges in the framework of multilateralism.

目 录

Contents

II　Keynote Speeches (Climate Change and Ecological Safety)

一、致辞与贺信

亚洲：追求更美好的明天

柬埔寨王国政府总理　洪森

尊敬的来宾，女士们、先生们：

十分荣幸能在第六届亚洲政党国际会议开幕式上致辞。

首先，我向所有出席亚洲政党国际会议与首届世界生态安全大会的所有尊贵的来宾表示热烈欢迎。我们在此欢聚一堂，继续探求亚洲政党国际会议的初衷，即推进亚洲各不同意识形态的政党之间的交流与合作，加强本地区人民和国家之间的相互理解和信任，通过政党的独特地位促进区域合作，从而在该地区创建一个可持续的、繁荣共享的环境。

快速发展的、日新月异的亚洲已经逐渐成为21世纪最有活力的地区。这体现在它快速的经济增长及其巨大的发展潜力。最近，它领导全球从有史以来最严重的金融和经济危机中恢复过来。然而，由于覆盖了若干个经济快速增长的、新型的工业化国家，该地区现在正面临着由于经济增长而引起的能源需求不断增加以及经济快速增长所带来的环境问题等巨大的挑战。因此，亚洲国家的领袖们在坚持可持续发展原则的同时，正试图在经济增长和环境保护问题之间寻求一种平衡。因此，我认为在此背景下，今天这个特别强调经济、能源和环境问题的大会主题——"亚洲：追求更美好的明天"是十分中肯的。我希望，此次大会以及相关会议能够有助于促进实际的合作和政治力量的参与，来实现我们共同的可持续发展目标，并找到解决能源需求问题和环境影响问题的方法。

各位来宾，女士们，先生们，我们今天举办这次大会并不是偶然的，这对于柬埔寨人民党来说是具有历史意义的一天。本次大会召开的日子，正是为了将柬埔寨从红色高棉政权中解放出来而建立的"柬埔寨救国民族团结阵线"成立32周年纪念日。柬埔寨人民党在过去的三十多年，经历了它历史上十分具有挑战的一段时期——将整个国家从毁灭和衰败转向发展和进步。

柬埔寨人民党非常自豪能够通过举办第六届亚洲政党国际会议，向世界展示：自红色高棉政权瓦解之后的三十多年来，柬埔寨在人民党的领导之下已获得重生，它正满怀着自信和坚定，稳步前进；而且，它的经济发展、制度能力以及政治自由也都得到了极大的加强。

毋庸置疑，在柬埔寨的发展过程中，柬埔寨人民党一直是主要的驱动力量。在过去的三十多年，柬埔寨人民党全心全力致力于造福人民。他们白手起家，在政治、社会和经济领域取得了不可胜数的重要成就。在柬埔寨人民党的"双赢政策"下，柬埔寨结束了红色高棉政权的统治，于1998年取得了完全的和平、政治稳定、国家统一和领土完整。另外，在柬埔寨人民党正确的经济政策下，柬埔寨取得了高速的经济增长，成为该地区经济快速增长的国家之一。在过去的十年中，GDP平均年增长约9.3%，这为消除贫困以及为人民谋取更好的福利和更高的生活水平提供了保证。

此外，至今我们能够取得如此大的进步，还要归功于我们对于区域及国际合作的重视。在长时间与外界隔绝之后，我们开始了与重要国际组织的关系正常化，并成为大部分区域及国际论坛的成员。如今，柬埔寨在国际舞台上正发挥着越来越重要的作用。我们正在着手进行维和军队的培训，为地区以及全球的和平作出自己的微薄贡献。1999年，柬埔寨成为东南亚国家联盟（东盟）成员国；2004年，柬埔寨成为最不发达国家中第一个加入世界贸易组织的国家。

各位来宾，女士们，先生们，我们今天聚集在此，对外界来说并不稀奇。这次会议只是该地区一次高端会议的其中一个，如最近的东盟会议、东盟+3会议以及G20峰会，这些会议致力于应对全球金融危机以及后续的恢复。应该强调的是，最近在韩国首尔举办的G20峰会显示，越来越多新增的政治和经济力量在国际政策的协调和全球管理中发挥着重要作用。

这次以"亚洲：追求更美好的明天"为主题的大会，是非常迎合需要和及时的，它正好在全球金融危机之后举办。由此我们可以发现，一个新生的富有活力的区域及全球经济框架正在形成——全球经济中心正在转向亚洲。这个趋势显示了重要的结构变化——通过该地区不断增长的投资、跨区域贸易和需求，将有越来越多的机会不断出现，但同时也伴随着能源和环境问题的挑战。

毫无疑问，维护世界和平，保持政治稳定，平衡经济和社会发展，迫切需要我们在地区、跨地区以及全球水平上的共同努力。然而，亚洲正逐渐崛起为一个能够引导世界经济从全球危机中走出来的地区，这是一个非同寻常的任务。我们相对其他地区而言，能够更快地通过刺激内需以及鼓励出口来恢复经济；在出口快速好转、健康的个人需求增长形势以及扩张性财政和货币政策措施下，亚洲经济在2010年的增长预计能达到8.2%。

在这种形势下，亚洲必须抓住机遇，及时解决区域的以及全球共同面临的挑战。亚洲在世界舞台上作用的加强，必须从重新评估可持续经济增长、能源安全以及人居环境开始，我认为这些是可持续和包容性发展的主要因素。

可持续经济增长

我想强调一个事实，亚洲未来的经济前景将取决于该地区如何成功地使各国资源在拉动经济增长中起到更加重要的作用。重新平衡地区内以及全球各个国家之间的经济发展，需要亚洲各国的共同努力。

对于亚洲地区来说，要减少依赖，扩展经济增长基础，就需要多样化地从出口带动增长中恢复的驱动力，使服务业和个人消费发挥更大的作用。这就要求我们拉动内需，通过以回报面向国内行业的投资为目的的结构改革来发展区域贸易，并通过强化商业气氛来消除障碍和瓶颈。

缩小该地区各国经济发展差距，对于区域竞争力和恢复力的具体化是非常必要的。亚洲强大的经济表现主要由印度和中国的经济增长以及韩国的强劲复苏显著引导；该地区作为一个整体来看，还显示着普遍的弱势。因此，我们需要通过加强亚洲经济平等以及增加利益份额——比如推进一体化、自由贸易和连通性——来确保包容性增长。像柬埔寨一样的发展中国家，其主要优先权应放在基础设施和人力资本的投资上。然而，这些投资需要大量的财政资源。因此，当前的迫切问题是改善流动性管理以及财政资源库，从而使亚洲经济对国家发展和地区发展有更多的自主控制。

的确，东盟成员国对于消除该地区各国之间的发展差异已经表现了强烈的决心。东盟持续不断地强化深入区域经济一体化，保持开放的地方主义；比如，清迈多边化协议、东盟连结蓝图以及东盟综合投资协议，这些都是为了加强区域投资，吸引更多外来投资，维持资本可持续性，增强弹性以及联接关键运输网络。另外，东盟+3自由贸易协议和东盟+6已经为贸易便捷化提供了参考模式。亚洲各国之间的进一步合作是经济合理的，将会促进这个世纪成为名副其实的"亚洲世纪"。因此，我认为很重要的是，在亚洲各国之间形成一个强大的经济联接（包括双边和区域贸易安排），主动深化区域一体化和多边贸易自由。这些合作不仅能促进经济发展，同时也为我们提供一个解决能源安全和全球变暖等其他类似问题的平台。

环境的可持续性

气候变化也在全球水平上对可持续发展构成了严重挑战。毫无疑问，全球变暖将会在程度和速度上继续增长。我们需要认识到，造成全球变暖的原因主要是人类活动。研究表明，某些人类活动（比如矿石燃料的燃烧和森林的不断砍伐）造成温室气体浓度不断增加，从而导致气候变化的产生。因此，迫切需要找到一种方法来应对气候变化带来的挑战，抑制温室气体的排放。"绿色经济"必须置于我们经济发展议程优先顺序的首位。我们还必须确保发展政策的各个方面都考虑了气候变化，以促进自然资源的可持续管理。同时，我们还需要增强对环境的保护。生物多样性的丧失是经济发展中显著的危险因素，也对长远的经济可持续性构成威胁。生物多样性并不是一种奢侈，而是确保全球经济增长和发展的战略财富。

发展中国家和发达国家要达成共识，共同合作，寻找解决气候变化的方案。建设性的咨询、合作和各国间了解的加深，毫无疑问会对应付全球变暖有所帮助。柬埔寨政府全力支持解决气候问题，但必须在联合国气候变化会议框架和所定的重要原则下，即"共同但有区别责任和各尽其责"原则，"发展中国家(特别是那些最容易受气候变化影响的国家)的独特需要和独特情况"和"预警原则"。所以，柬埔寨政府要强调发展中国家应对气候变化的努力与发达国家财政支持和技术转让的关系。

比如，对湄公河地区国家来说，可持续发展是指在社会经济发展和环境保护之间找到一个适当的平衡点。越来越多有关环境保护的计划、项目和研究如"十年迈向绿色湄公河区域"计划和"鸠山倡议"已经展开，解决共同的环境问题，如森林砍伐、森林和土地退化，以及由于经济快速发展和人口增长所带来的空气和水污染。然而，湄公河区域国家需要财政援助、技术转让和足够的设备保证，来发展清洁能源、改善生物多样性、管理渔林、发展污染控制管理装备。

能源安全

我们都很清楚，未来的经济发展会导致能源需求的增加。虽然政府努力推行再生科技，但在未来的一个世纪，为了发展经济，全球对化石燃料(如石油、煤和天然气)仍然十分依赖。研究估计，在2050年全球能源需求会加倍。为了应付快速的经济发展，亚洲对商业能源如石油、煤和天然气会日益依赖。比如，中国将在五年内成为世界最大的主要能源使用者，而印度会继中国之后成为世界第二大主要能源使用者。预计石油和天然气价格将进一步明显上涨。所以，我们必须大力提倡使用和发展再生能源，因为能源安全对可持续经济增长和未来发展十分重要。

此外，不管对这个地区还是所有的国家而言，我想相互合作确保能源安全是必要的。与此同时，这种共同合作对亚洲来说也是有益的，因为很多亚洲国家都是世界上主要的能源消费者和进口国。通过提高对亚洲地区的投资比例，东亚有巨大潜力可成为绿色技术的前沿。

我们要了解发展中国家的发展需要，特别是最不发达国家要在应对气候变化和能源需求的紧急性中寻找一个平衡。事实上，改进技术效率和更好地利用替代和可再生能源也是解决能源安全与气候变化的关键。所以我们必须要鼓励增加对清洁能源发展机制的投资。

提高能源利用效率亦十分重要。更清洁的能源科技（先进的化石燃料和再生能源科技）也同样重要，它们可以创造就业机会，促进工业发展，减少空气污染和帮助减少温室气体排放。提高能源利用效率不但对能源安全有好处，亦是建立环境可持续性发展的必要因素。再者，核能应该是一个可行的选择，因为核能是一个重要的无碳能源的来源。不将核能纳入其中的一个可行方案，是不利于长远控制二氧化碳排放的。

地区主义：湄公河地区国家

地区主义日益重要。这反映在全球的自由贸易协定迅速增长和对多哈会谈进程缓慢的不满上。所以我想与大家分享关于湄公河区域的最新进展，湄公河区域合作对亚洲区域合作的建立起着重要的作用。

湄公河地区的发展对东盟和东亚地区是十分重要。因为它会缩小东盟国家之间发展的差距，并成为湄公河地区经济一体化，确保和平与稳定，以及在东盟与东亚建立一个可持续发展与和谐社区的先决条件。

当然，基础设施建设的发展在东西经济走廊和南方经济走廊项目中的目的是联系主要的交通网络和湄公河地区的跨国贸易，将对经济发展、减低发展差距和推广地区广泛一体化作出贡献。到目前为止，这些发展工作未能完全地开展，特别是基础建设，如南方经济走廊项目的桥梁和道路，所以我想呼吁发达国家和合作伙伴对此更加关注，以实现这个目标。由于这些项目需要大量投资，所以及时的财政支持是必须的。

再者，生活在全球一体化的时代，由于对地区一体化重要性的充分理解，所有湄公河区国家都有职责对湄公河作为一个整体对其环境进行保护。在这方面，为了维持区内和谐和繁荣，所有湄公河国家应尊重彼此的利益，考虑当项目实行时对其他国家造成的影响，以及确保所实行的项目应该对其他湄公河国家的生态环境是无害的。

女性参与和青年领导

利用这个机会，我想强调在发展过程中女性参与和青年领导的重要性。长期的可持续发展和亚洲的未来在于下一代领导人，特别是亚洲国家年轻的领导人。在培养下一代领导人方面，亚洲国家必须起到至关重要的作用，要通过加强并扩大教育、就业和培训计划，鼓励更广泛的参与。此外，如果没有女性重要的贡献和参与，社会、经济和政治就不能平稳发展。所以，我们需要积极与女性合作。如果没有女性的广泛参与，亚洲革新的过程就不能实现。

与此同样重要的是，我们还应加强文化合作和交流。学习和提高对每一个国家的文化、风俗和礼仪的认知会对相互尊重主权和对文化遗产保护作出贡献。每一个文化都是独一无二的，文化遗产对民族认同感尤其重要。所以，我认为，文化体现我们是谁，是表达我们的风俗和传统的强而有力的方法。所以我们拥护多元文化，承认每一种文化的价值观和重要性。

此外，为进一步鼓励和加强了解，以及深化不同文化间的知识和信息交流，建立文化协会是适当的和合乎逻辑的。这些协会不但帮助传播信息和教育人民关于不同文化有趣的一面，亦对为文化保护作出贡献的人表达敬意。

在致辞结束前，我想利用这个机会对这次会议的组织者表示深深的感谢，包括亚洲政党国际会议秘书处，相关部门和机构。正是由于他们的努力工作，会议才能在今天顺利举行。

我再次代表柬埔寨人民党、柬埔寨王国政府热忱欢迎各位使节、女士们、先生们参加在柬埔寨举行的第六届亚洲政党国际会议。我希望这次会议能为亚洲更好的明天提供最好的发展路径。我想以佛教的四个祝福：长寿、福德、健康和力量，来祝愿各位使节、女士们和先生们。

最后，让我宣布第六届亚洲政党国际会议开幕！

加强政党对话，促进亚洲和平与稳定

亚洲政党国际会议创始主席　何塞•德贝内西亚

第六届亚洲政党国际会议主席洪森总理阁下，

马来西亚总理阁下，

尼泊尔总理阁下，

印度尼西亚前总统梅加瓦蒂·苏加诺阁下，

菲律宾前总统拉莫斯阁下，

印度尼西亚副总统尤素夫·卡拉阁下，

蒙古前总理阁下，

中国共产党李源潮部长阁下，

亚洲政党国际会议联合主席郑义溶阁下，

女士们，先生们：

我注意到，我们的一些代表偕夫人一同光临，正如我一样。我记得去年在欧洲的政党会议上，我问政党会议的瑞典籍主席："为什么您在所有会议上，总是带着您的妻子？"他对我说："德贝内西亚先生，我总是带着我的妻子，因为我总是带着我的政府。"于是，我回答说，"是的，您带着您的政府。但对我而言，我带着我的妻子吉娜，是因为她是我的'执政党'。"

各位阁下，接下来我想向洪森阁下、索安副总理和柬埔寨的政治领袖们、亲王和国王，为他们卓越的领导才能致以诚挚的感谢；当然还有奉辛比克党的卓越领导，他们的努力使本届大会得以顺利召开。

如您所知，洪森总理尽了最大努力把柬埔寨的政治力量聚集在一起处理发生在这里的危机，我感谢总理阁下对于那些上周在金边事件中遇难者和伤者家属的关注。

不仅是我们的洪森总理，柬埔寨的其他政治领导也发现柬埔寨人民团结起来后的巨大力量，经历内战的洗礼之后，他们完全有资格为祖国而献身。

一个小时以前，洪森总理告诉我们，在20世纪80年代当他们从丛林中走出来的时候，当时只有七个人在金边，他召集他们一起，留在这里，以便把这个国家精心治理成为一个充满活力的民主国家。

祝贺您！洪森总理！

洪森总理、索安副总理，还有其他领导人与联合国一起，极具勇气地依法惩处了那些对柬埔寨人民犯下大屠杀滔天罪行的刽子手。

再一次祝贺您！洪森总理！

我还要向王室、亲王和国王表示祝贺，他们以一种其他国家无法做到的恰到好处的方式使柬埔寨恢复了稳定，他们有能力做到，并有效缓和了柬埔寨与最近的邻国的关系，如越南和泰国。

近些年来，政治稳定使柬埔寨的经济无论是在区域还是全球领域都有了很大扩展。同时我们也见证了柬埔寨的经济潜力，它是如此快速地实现了经济复苏。

新的柬埔寨反映了当前我们这个时代的精神，柬埔寨为亚太地区的经济发展贡献了力量，而亚洲也将成为这一趋势的领导者。

21世纪一定是亚洲的世纪，当然，今天在这里出席会议的女性可能会说："是的，这是亚洲的世纪，但21世纪同样也将是女性们的世纪。"在亚洲政党国际会议的各位代表面前，我无法掩饰对亚洲未来的乐观心情。

10年前，我们有来自21个国家的大概50个政党聚集在马尼拉，那似乎是我们的极限了。但现在我们的会员急剧增加，至今我们已经拥有来自82个国家的318个执政党和参政党，他们来自东北亚、东南亚、南亚、中亚、西亚以及阿拉伯世界，同我们一起参与亚洲政党国际会议。我们的大会迅速发展，2000年在马尼拉，2002年曼谷，2004年在北京，2006年在首尔，2009年在巴基斯坦的阿斯塔纳，同时这些国家也在扩展常务委员会成员。

更重要的是我们已经建立起了一个相互理解和相互扶持的网络，致力于亚洲政党之间的合作，无论是执政党还是参政党。

通过亚洲政党国际会议，我们的许多成员都在为国家的繁荣而努力，并有了一种最终可能超越意识形态和地域差异的合作意识和共同发展的工作。实现这一切当然需要亚洲政党国际会议的大力支持，而这种支持必须来自各亚洲政党。因此，我相信，正如亚洲政党国际会议吸引力不断增加一样，我们的工作也为共同加强政党制度发挥了很大作用。坚强稳定的政党制度能够为我们带来长久稳定的政治局势，为政治网络带来转变，并以更持久的政治联盟取代个性化的权力关系。

如各位所知，我们也开始了和拉丁美洲、加勒比地区和非洲的政党寻求合作。我们的常务委员会已开始和拉丁美洲和加勒比政党常设会议（COPPPAL）的干事们定期会面，并首次参加了在阿根廷布宜诺斯艾利斯举行的历史性的会晤，以解决我们的共同问题，特别显示了我们在面对全球气候变化局势时团结起来的力量。墨西哥政治领导人们这次也来参加我们的会议，让我们用热烈的掌声向他们致以问候。

同样，我们与非洲政党也进行顺畅的沟通与合作。最初是通过非洲联盟，两个星期前，我们在肯尼亚举行会议，建立了跨亚洲、拉丁美洲和非洲的三方合作关系。我们将三个发展中大陆联系在一起的目标最终通过政党得以实现，很快我们会和在欧洲和北美的兄弟和朋友们建立起联系。所以有朝一日，这个联合将能传达一个全球性的政党的声音。

在区域和全球层面上，亚洲政党国际会议已经开始表达亚洲的观点。作为亚洲政党国际会议的创始人，我们感谢有60个国家为我们举办大会提供赞助。我们一些领导人正在考虑马来西亚提出的关于提供资金、消除贫困以及应对气候变化的方案。与此同时，联合国秘书长也在为我们建立全球性的消除贫穷基金的提议进行倡议，其他的小额贷款机构可以从中申请资金。联合国也考虑将亚洲政党国际会议作为联合国大会观察员提上日程，但我们需要来自各国政府的支持，请最晚于明年向联合国提交申请，越早越好。

当我们开始第二个十年，我们也进入了一个"党对党外交"时代，我们已经与我们的两个主要成员印度国大党和巴基斯坦人民党进行了沟通，我们也在竭尽所能寻求缓和这两个国家间历史、种族和宗教方面问题的方法，正如我们的常务委员会所提议的那样。洪森总理昨天与我们谈话并坚决支持设立和平委员会，这个建议也是为了两国的利益，昨天已经在亚洲政党国际会议通过。

东北亚半岛也将开始探索以一个类似亚洲政党国际会议的形式把韩国和朝鲜政党撮合起来的可能性。让我们支持韩国的政党和朝鲜劳动党，并帮助他们创造政党外交的开始。

这种模式与亚洲地区其他冲突相关。以此为背景，亚洲政党国际会议支持并参加亚太中间党派民主国际，以努力通过在冲突地区加强政党间的对话，促进亚洲和平与稳定，例如：尼泊尔和阿富汗。

亚洲政党国际会议根据尼泊尔三个主要党派的邀请，将派代表团去加德满都，寻找三方都关心的，并通过对话打破僵局的道路。

亚洲政党国际会议也欢迎阿富汗政府设立国家和平委员会，目的是与暴乱者开始谈判，以政治途径平息冲突。

亚洲和世界和平必将继续发挥其带来改变的重要作用，作为国际上的政治、宗教和种族国家，我们必须保护弱势阶级。

我们还支持中国、越南、菲律宾以及东盟各国在南中国海问题上所做的回应，我们感谢所有中国、日本还有东盟的政党，在我们处理南中国海的问题时，也是奉行这一"党对党外交"的精神。

　　我们还必须注意到全球化背景下新的世界经济和政治秩序：我们正在发展一种共同的思想观念体系，以服务于世界和平这一主题。我坚信，该系统比起公营和私营合作伙伴关系，它可以使亚洲处于发展阶段的私营企业更好地携手合作。洪森总理今天谈到要建立、扩大和深化本世纪的政党体系，以使我们可以接触到左倾力量和右倾力量，这样我们或许可以选择比较中和的道路，也许这就是世界各政党和各意识形态之间通过一定程度的妥协所能达成的共识。

　　回忆起在2000年9月份马尼拉大会常委会上，我那时就注意到我们已经聚集在一起，将在亚洲的政治历史上写下了自己的新篇章，我们共同的愿望就是，在所有聚集在这里的民族之间，将共同建设和平，建设自由，建设繁荣。

　　我们见证了亚洲的经济和政治一体化的阶段目标。十年前，这个目标似乎是乌托邦，但我们子孙后代将见证其实现。我们会像老人一样，种植树木让我们的子孙乘凉。我们面向亚洲的未来采取行动。现在，亚洲一体化已成为一种可能的梦想。对你对我来说都有理由为亚洲政党国际会议的第一个十年感到高兴，并为亚洲政党第二个十年做好准备。

　　在我们继续推进亚洲政治、经济一体化时，我们所见证到的这种趋势虽然缓慢但已经清晰地出现，首先是东南亚地区的东南亚国家联盟，在湄公河流域，东北亚，亚洲议会大会，南亚的区域合作南亚协会，亚太经合组织，俄罗斯、中国和中亚区域的上海合作组织，阿拉伯国家合作会议组织，海湾合作委员会，南亚次区域的经济合作组织以及其他伟大的组织。是的，这个趋势是缓慢的，但已经清晰可见。作为亚洲人，我们说，有一天，会有一个亚洲共同体，有一天，会有一个亚洲联盟。

　　谢谢！

加强政党对话有助于区域合作

联合国秘书长　潘基文（视频致辞）

柬埔寨王国政府总理洪森阁下，
尊敬的各位政党领导人，
女士们，先生们：

首先，请允许我就上个月送水节的悲惨事件向柬埔寨王国政府表达慰问，在这个悲伤的时刻，我的心是与所有柬埔寨人连在一起的。

女士们，先生们，我向亚洲政党国际会议十周年表示祝贺，并向其一直以来致力于促进亚洲地区的对话与合作所取得的成就表示祝贺。各个政党可能有截然不同的关注点、背景和意识形态，但是通过经常性的沟通和交流，政党将准备好促进相互理解，获得更多的民主以及加强区域合作。更重要的是，他们有责任这样做，并且无论他们是否掌权，都应被允许这样做。

这是考验的时代。世界正面临着诸多挑战，没有一个国家能够独自应对，例如，气候变化、生态恶化、极端贫困、经济危机的持续影响、核威胁、自然和人为灾害、军队暴力和对人权的侵犯等。世界各地的人们生活在日益严重的焦虑和恐惧中。然而，在联合国为解决这些问题（包括为实现千年发展目标）所做的努力工作中，亚洲政党国际会议都是一个强有力的支持。

同时，我要感谢亚洲政党国际会议为促进女性在政党和社会活动中的参与所做的努力，并感谢其帮助青年人在政治决策中发挥更大的作用。我们的工作应该尽量关注于我们如何携手合作，并使联合国能够参与进来。

再次感谢你们对联合国的支持，祝愿会议取得圆满成功！

政党需要顺应时代的改变而改变

马来西亚总理　纳吉布·敦·拉扎克

首先，请允许我对于亚洲政党国际会议（ICAPP）邀请我参加第六届亚洲政党国际会议和首届世界生态安全大会表示衷心感谢。这对于我来说，是非常有意义的一件事情。我很高兴今年能够第二次来到这里，再次拜访柬埔寨金边这个美丽的城市。同时我也想对洪森总理阁下以及柬埔寨政府的热情款待表示感谢。

对于送水节时金边大桥上由于踩踏造成的重大无辜伤亡，我与之前的演讲者一样，感到十分悲痛。我谨代表马来西亚人民和政府，向柬埔寨政府和人民转达我们对于这个不幸事件的深切哀悼。

同时，热烈祝贺ICAPP创始主席、ICAPP理事会联合主席何塞·德贝内西亚阁下建立了这个成功的论坛，将亚洲各国政党聚集在一起，促进他们之间的交流合作，从而取得更大的区域性谅解和凝聚力。今天我非常荣幸能够站在这里，出席亚洲政党国际会议十周年大会，十年前，即2000年9月，这个会议曾在马尼拉召开。我也很高兴能够作为政府首脑，而且代表我们国家执政联盟出席此次大会。令人振奋的是，今年有来自51个国家和地区的317个亚洲政党参加了此次亚洲政党国际会议。当然，来自各个不同政党的丰富主张和经验将会形成有趣的思想碰撞。

本次ICAPP大会的主题"亚洲：追求更美好的明天"是十分及时的，反映了亚洲正在持续转变成为世界上一个高速增长的地区。全球化的趋势以及中国和印度作为经济主宰力量的崛起，改变了世界对亚洲的看法，也改变了亚洲对自己的定位。随着世界观和政治观的改变，社会态度和思维倾向也随之变化。在信息系统技术风靡全球的大势下，这种变化更加剧烈。忽然之间，大街上的每个人都有机会对那些一度只是政府和官方独有领域的事件发表自己的观点和评论。忽然之间，社会评论的空间大大开放，这个空间被运用、甚至超越公德的限制和法律的界限被滥用。我经常说，当今的政治，已经不再是我们所说的一如往常，而是非同寻常。

鉴于此，没有任何一个有头脑的政党敢忽略这些变化，因为这些变化影响着的正是这些政党所服务的人们。无论一个政党是代表传统观念还是代表无拘无束的现代主义；无论他们的存在是为某个特定意识形态还是某个目标而奋斗，这些影响着整个社会的变化都重要到无法忽略。整个政治版图都发生了如此剧烈的改变，不管我们喜欢与否，我们都必须有所认识，从而使政党采取相应的行动。

纵览近代历史，我们可以发现许多政党都考虑到了动荡的政治民主，这种政治民主在他们统治的巅峰时期看似无懈可击，却往往由于他们没有顺应时代的变化而导致命运逆转，最终走向灭亡。人们经常说，权力是一种兴奋剂，能导致很多问题，比如惰性和健忘症，甚至引起幻觉。一个长期掌权的政党往往遭受这些问题的折磨：惰性，就是说止步不前，安于现状不愿改变，变得仅仅满足于目前所取得的成就；健忘症，即它忘记了它成立的初衷和它本应实现的目标；幻觉，指的是幻想它的政治支持基础是永久不变的，因此误以为它将永远大权在握。一个政党不会仅仅因为它为这个国家取得了自由就会永远当权，除非这个政党自己能顺应时代的改变而改变。

许多政党都遭受着这些问题的折磨，因为它导致党派层级制度下的内部争斗以及权力斗争，而这些通常会无可挽回地毁灭这个政党。最终结果就是失去选民的信任和政党的民众基础，从而使得民众转为支持能够更好地满足他们需要的其他政党力量。最后，曾经看似无懈可击的政党忽然之间发现它崩溃于自己内部斗争的重压和弊病。到那时候，除非有奇迹发生，否则它将无法扭转其衰败的命运。

我今天提出这一点，是因为我感到对于任何一个政党来说，意识到自己需要改变是非常关键的。并不是为了改变而改变，而是为了更好地满足这个政党的支持者的需要；以更高的效力和相关性为目的进行改革，为了与时俱进而再创造。与本次大会的主题"亚洲：追求更美好的明天"相对应地，政党们必须开始认识到，一个更加美好的明天是建立在该地区一个更好的政治环境基础之上的。

朝着这个终点努力的首要任务，就是加强政治过程。必须加大与公众接触的力度，了解他们的需求，以便政党能够更好地为其人民服务。事实上，当前的趋势表明，社会大众对于参与政治过程和国家建设有着迫切的需求。一个理解了"这些就是政治活动的新规则"的政党，就是未来成功的政党。

在马来西亚，执政联盟认识到需要改变我们为我们的支持者——主要是一般大众服务的方式。我们有顺利的时候，也有颠簸而行的时候，比如说在2008年3月大选中失利。这在当时的确为我们敲响了警钟，但是从那时起两年以来，尤其是最近的这一年半以来，我们的执政党总的来说经历了重大的变革，而且，由13个政党组成的执政联盟在改革以及重新取得人民的拥护等方面都突飞猛进。

如今，政治联盟国民阵线——我担任主席——持续地创新自己以适应时代的变化。最近，我们通过了一项修正案来认可直接进入联盟，从而扩大我们的支持基础，允许任何个人或组织，无需正式加入联盟中13个党派的某一个，直接进入执政联盟。马上，很多非政府组织和非营利组织就表达了他们与我们一起合作，为国家更大的进步和更好的发展共同努力的强烈愿望。

在我担任主席的马来民族统一机构（UMNO），2009年通过了具有历史意义的党章修正案，允许各级政党成员更多地参与到执政党领导班子的选举当中。当然，这使我们的工作得到更多的检验，也将很多政党领袖包括我置于大量的政治风险之中。但我们相信，我们的政治支持应该也必须取决于我们是否具备做好的能力，是否具备说服人们现在支持我们的能力。在这种机制下，我们将会由我们自己的政党成员亲自来判断我们是否胜任他们的领袖。

除了内部调整以适应时代变化，对于一个政党来说，最重要的事情，正如我之前提到的，就是一贯而持续地与人民联系在一起。这一点，正是人民决定一个政党是否与他们在一起时首先考虑的一个要素。类似他们的下一份薪水从哪里来，他们怎么才能吃好喝好，他们怎么支付孩子的教育费用这样的事情，对于一般大众来说才是极其重要的事情。他们对于可能弄坏他们汽车的附近大街上的大坑和可能影响他们上学上班的附近河上坏了的大桥的担心，要远远多过他们对于政治圈中争权夺利的关心。

我一贯坚持政党必须与人民的命运相协调，他们必须随时准备满足社会的期望，因为如今的社会信息远比以前灵通，它能与远近其他社会进行对比。正因为如此，2009年4月我就职时第一个议程就是将"以人为本"、"现在就行动"置于马来西亚大旗下。它已不仅成为我们的国家口号，而且成为我们的国家理念。我们要确保我们交给人民的是重要的东西，而且我们要让人们知道我们的所作所为。从那时起，我们已经采取了许多措施，比如，建立马来西亚医疗服务从而给农村地区提供免费医疗援助，设立马来西亚奖学金来奖励优秀的学生——不论种族和宗教，我们还加大力度推出了各种方案，计划在全国范围内改善低收入群体的生活条件以及消除贫困。

在经济方面，国民阵线政府正在不懈努力，将马来西亚转变成为一个高收入经济体。2010年1月，我们公布了"政府改革规划"（GTP），并确认了6个国家关键成果区（NKRAs），即减少犯罪、反腐败、提升低收入群体生活水平、改善农村基本设施建设以及加强城市公共交通。所有这些都是对大众人民的生活造成直接影响的重要区域。不久后，经济改革规划被公布，强调建立一个高收入的、持续的、包容性的国家，并确认了12个国家关键经济枢纽区，这些区域将给予特别关注，以进一步加快马来西亚前进的步伐。这里需要强调的一点是，实际上，由于这些措施将对民众的日常生活造成影响，民众得以参与到公众事务的试验和讨论中来，因此在准备这些经济措施的每个阶段，我们都能够通过他们的参与来推测他们的预期并满足他们的愿望。

当然，对于我们是否满足了民众期望的评价只能在下次大选时才能知道结果，但是，据目前显示，民众给予了正面回应，而且正在重建他们对于国民阵线的信任。由于我们开始实现竞选宣言和我们曾许下的承诺，在民众眼里，我们的可信度正在进一步加强。此外，马来西亚的人民将马来西亚执政联盟视为一个经得起考验的、不同政党朝着共同目标奋斗的真正合作伙伴，而不是仅仅为了政治权宜之计草率组合起来的意识形态相互对立的联合体。

当我谈到亚洲的更加美好的明天时，我会小心不触及外部政治环境。亚洲是一个经济飞速发展的地区，是全球贸易的关键一环，承受不起极端主义或恐怖主义带来的骚乱和破坏。在这种背景下亚洲政党必须发挥重要的作用。

当我在9月份的联合国大会上发言时，我呼吁全球的中立派行动起来，重新确立我们失去的中心点和战略高地，我呼吁中立派排斥那些因为一己之固执和偏见而威胁整个世界的极端主义者和恐怖主义者。在这一点上，没有任何地方比在亚洲更重要，亚洲的一些地方极端主义分子还在持续地传播他们的仇恨。

作为亚洲这个大家庭中的一员，我们大家都有权利和义务确保这个地区的自由和安全，远离意识形态的冲突、破坏、不团结和任何非正当原因的恶意。极端主义分子总是企图在沙子中划线，划分出互相对立的两方，产生实际上并不存在的惩罚范围。我曾经不止一次地说过，真正的问题并不在穆斯林和非穆斯林之间，而在极端主义者和信仰伊斯兰教、基督教、犹太教、印度教或任何其他宗教的中立派之间。

我们决不允许极端主义者和恐怖主义者扰乱我们的社会秩序，决定我们各个国家对话的方向。我们必须推行理性达成共识。中立派必须是唯一的占统治地位的声音，互相合作和协商必须是优先选择的途径，而不是敌意和对抗。

从这一点来说，政党成为了关键的一环，政党的自然属性是体现其特定的理想、奋斗或目标，而且它们的成员将为这个奋斗目标而不遗余力地保护、宣传和支持。不幸的是，有时其中一些政党或政治派系为了一时的政治利益会走上极端主义的歧途来巩固它们的地位。这样一来，为了获取短期的政治利益，它们将激起怨恨，点燃熊熊怒火。这个时候，中心位置被夺走，愤怒和危险氛围激发了不和谐的声音。同时，该政治派系的反对方将发出极端主义的声音来反对它们对手的言辞，从那时起，这个国家将在很短时间内陷入完全的混乱。

　　不管出于何种原因，政党都必须远离极端主义的歧途，解决冲突总有和平的方式，任何短期的政治利益都不值得牺牲国家和地区的和平与和谐。最终，最重要的是人民的福祉和对我们价值观、文化、生活方式的保护。我呼吁亚洲的政党们加入到我们全球中立派的动员中来，反对政治仇恨和任何形式的极端主义。

　　的确，亚洲政党不需要依靠极端主义者故作姿态来维持相关性和受欢迎程度。相反的，在如今这个信息技术的文明时代，正如我之前说过的，最受欢迎的政党应该是那些最与时俱进并能顺应人民的愿望和需求的政党。极端主义者依靠这些恐惧的和看似受威胁的人或事得以存在。如果我们让大众明白，担心是没有必要的、隐患是不存在的，那么极端主义者便不复存在。

　　在呼吁全球中立派运动过程中，我必须强调联系和网络沟通对于推进实现中立主义理想的重要性。除了双边及多边的政府联系之外，一些类似在论坛上建立的非正式沟通网络，对于向特定的听众传达正确的信息将起到关键性作用。我强烈建议，所有在场的亚洲政党建立并加强互相之间的联系，分享知识和经验，然后将这个共同的信息——希冀一个更加美好的未来传达回去给你们的人民。你们都是你们社会的领袖，因此你们处在这个特殊的位置上向整个世界传播这个关键信息——中立主义。如果我们能够把这个信息带回家、带给我们的人民和我们的社会，它将成为人民生活的信条，那么极端主义将会真正地衰败，真正的中立派运动将会产生效果，不仅仅在领导者水平，而且将惠及一般民众。

　　我已经和你们分享了我对于亚洲政党如何在本国国内和外部政治环境背景下前进的一些想法。同时我还与大家分享了马来西亚的一些经验，虽然它肯定不是完美的，但是我希望我列举的这些例子能够引起大家的深思。

　　每次拜访柬埔寨，我都禁不住回想20世纪90年代初军国政府时期我作为国防部长第一次来柬埔寨的情形。那个时候，柬埔寨的未来十分不确定，我们根本无法想象如今所看到的柬埔寨的样子。从30年前到现在的柬埔寨，这又是一个国家的成功史。而这一转变，要归功于洪森总理的领导力，他将柬埔寨从绝望中拯救出来，成为一个稳定的、进步的国家。很多人问我是什么使得一个国家是成功的而其他国家不是？主要区别在哪里？答案最终都归结为一个词："领导力"。大胆的、有勇气的、有远见的、实际有效的领导力，这才是关键，而洪森总理具备这种领导力。

　　如果所有的亚洲国家都能如此，那么21世纪将会成为真正的"亚洲世纪"。

在政党网络中共享经验和看法

尼泊尔民主共和国总理   马德哈夫·库玛·尼帕尔

尊敬的柬埔寨王国政府总理洪森阁下，
亚洲政党国际会议组委会主席、柬埔寨副总理索安阁下，
亚洲政党国际会议常务委员会的联合主席们，
各国代表团团长阁下，
各位尊敬的部长及政党领袖，
各位尊贵的客人和各位同事：

我们在这里举办第六届亚洲政党国际会议，这个会议是我们在新世纪之初所创造的。在这次大会开始以前，我对11天前发生的柬埔寨国家悲剧深感悲痛，当他们快乐地庆祝一个最受欢迎的本地节日时，发生的踩踏事件导致很多人失去了宝贵的生命。我想对丧失亲人的家庭表达我们深切的哀悼，同时也很欣赏柬埔寨政府以平静和冷静的态度处理这个重大的民族悲剧。

回想起亚洲政党国际会议的十年历史，我感到巨大的喜悦。它已经能够吸引并在亚太地区建立一个日益广泛的政治党派网络论坛。前十年对于亚洲政党国际会议组织并巩固自己成为独特的政治论坛是至关重要的。第六届亚洲政党国际会议的主题是"亚洲：追求更美好的明天"，它真实地反映应了我们在日益增长的活力背景下的对和平与发展的集体承诺。在此背景下，亚洲地区近年来在社会经济方面取得了显著的进步。同样值得注意的是，鉴于女性和青年在国家政治生活中的重要性，女性政治家专题会议和青年政治领袖专题会议这两个研讨会也得以举行。

当我回首过去，想起我们从2000年9月的马尼拉开始我们的旅程，通过泰国（2002年）、中国（2004年）、韩国（2006年）和哈萨克斯坦（2009年）逐渐向前迈进，直到今天抵达金边举行的第六届亚洲政党国际会议，我感到很满意。当我们向前迈进并巩固我们的成就的时候，我的思绪还是回到了形成期的2000年。那是我第一次见到何塞·德贝内西亚阁下，并与他交流意见。这次会面使我们的想法与思想更加具体化，并最终促成了亚洲政党国际会议的成立。

今天，我很骄傲地回忆起这历史性的起点，我们从那里一起开始了这个漫长的旅程。

我很高兴亚洲政党国际会议不再局限于亚太地区，因为我们已经与拉美和加勒比政党常设会议的协调处建立了联系。他们在这里与亚洲政党国际会议常务委员会举行第二次联席会议。（第一次于2009年7月在布宜诺斯艾利斯举行。）亚洲政党国际会议与拉美和加勒比政党常设会议一起包含了68%的世界人口，53%的全球国内生产总值和50%的全球大陆面积。亚洲政党国际会议怀着同样的热情与非洲建立发展合作关系，推动全球和平与繁荣。作为一个及时的和日益政治化的论坛，亚洲政党国际会议变得越来越受欢迎。我相信，这要归功于由于亚洲政党国际会议领导人高昂的情绪、奉献精神和归属感，他们广泛定义和描述了亚洲政党国际会议的核心价值观和前景。

当我们开始我们的使命时，我们希望建立一个政治网络，共享经验和看法，从彼此的成功中学习良好的管理、以人为本的发展、法治、透明度、公共责任、腐败控制、民主体制建设和减少贫困。其他相关问题的广泛的讨论和商议已经在亚洲政党的支持下举行，包括在这时期秘书处与联合国的接洽。

作为公众动员和提高普通百姓福利的公认的民主工具，政党要在启发和授权人民建立一个公正、平等和繁荣的人类社会方面发挥重要作用。我们要在这样的背景下理解和评价政治领袖们的角色和责任的重要性。亚洲政党国际会议吸引了各种政党——左派、右派、中间派——它们来自南亚、东南亚、东亚、西亚、中东和太平洋。这使亚洲政党国际会议成为一个真正的广泛的国际政治论坛。可以理解的是，政党领袖们的清廉形象和效率成为道德权威的来源。因此，由国家提供资金的政党会促进政党的透明度、责任心和基于价值观的政治。在过去十年里，我们在这些领域投注了很大的精力与努力，并取得了很大的进展。我相信，对于所有积极参与到此论坛的人来说，这是一个很好的学习经验。在不断变幻的国内国际背景下，我们需要应付无数的政治性及非政治性的挑战。我们相信亚洲政党国际会议会继续为我们提供一种有用的接触性交往及共同合作的平台，在该地区和其他地区改善人类生活和总体福利。

通过促进世界和平、正义和繁荣，亚洲政党国际会议能在差距不断扩大和经济边缘化的现代世界中发挥巨大作用。即使国家间相互依赖的气氛日益浓厚，当前全球化过程中大多数的发展中国家和最不发达国家仍然感觉到被排斥和剥夺。这些问题需要得到合理的解决。作为最不发达国家集团现任主席，尼泊尔特别关注着这些国家的穷人们的持续困境，他们每天都在为最低限度的生活必需品挣扎。

各位朋友们，容许我利用这个机会与大会分享我国的最新发展。尼泊尔在制度上进行了改变，现在正通过由选举产生的制宪会议拟定新的民主宪法，以使政治变革体制化。我们已经开始了各种和平协议，结束了长达十年的冲突，使叛军回归国家主流，制定了临时宪法，选举了最具包容性以及广泛性的制宪会议，结束了240年历史的君主政体，并选择以联邦制度代替以前的集权制。

我们仍然有两个主要的任务在手，一是完成和平进程，二是起草民主宪法制度化，使和平、稳定和发展制度化。我不遗余力地在我国捍卫民主、法治和全国和解。我们有六个月的时间写完宪法，联合国尼泊尔特派团还有一个半月的时间就会离开尼泊尔。我们要在他们离开前完成和平进程。尼泊尔人民、各方祝愿者及国际社会都渴望和平进程尽快完成，新宪法能在规定的时间内公布，社会和经济发展顺利、快速地进入轨道。

在我们努力达成和平进程，并颁布新宪法的同时，我们还致力于共同实现社会和经济的快速发展，这也是我国人民的愿望。我十分乐观，在国际社会的祝愿和我国人民的积极参与下，在不久的将来我们将建立一个自由和繁荣的国家，一个享受持久和平与兼容的多元主义民主、法治和联邦国家。

总理阁下，尼泊尔和柬埔寨之间的亲切友好关系给进一步加强各个领域合作带来了良好的前景，这些领域包括旅游、文化和考古等方面。作为佛祖降生圣地的兰毗尼为两国人民提供了强烈的共通感，我对此十分高兴。作为一个热爱和平的国家，尼泊尔在世界各地不断地传播和平的信息，同时也积极参与联合国的维和任务。在有关和平进程和国家和解方面，柬埔寨可能是一个最受瞩目的例子。它展示了一个国家尽管被内部纷争破坏到最大程度，依然能够通过强烈的政治决心恢复民族凝聚力、团结、稳定、和平与进步。尼泊尔期望从经验中学到一些有价值的东西。

尊敬的朋友们，我祝贺柬埔寨政府及人民举行这次重要会议。我同样希望祝贺新的第六届大会主席，希望他们成功地推进这个论坛，探索和推广政党、政府和这个地区国家间的新的合作途径。我们每个人都经验丰富。我们可以把我们的经验聚集在一起，并利用它们以达我们的共同目标。亚洲政党国际会议绝对有潜力和希望为我们服务。

我相信金边会议将会增添一个新的章节。通过确认新的合作领域，它将会进一步促进国家、政党和地区领导人之间更好的理解和富有成效的合作。多数参加亚洲政党国际会议的国家人口都是贫穷和营养不足的，并一直受变幻无常的气候变化影响。尼泊尔坚信我们会把扶贫并作为我们的重要目标。

谢谢！

21世纪召唤我们立即采取行动

菲律宾前总统　菲德尔·拉莫斯

洪森总理阁下，

各位政府首脑和国家元首：

感谢各位阁下参加此次亚洲政党国际会议，请允许我对大家表示热烈的欢迎，并祝愿所有来自亚洲的人们以及全世界的人们生活幸福，友情长存，身体健康，长命百岁，机遇多多，有一个更加美好的明天。

21世纪要求我们要少说多做。对于与会的每一个人来说，开幕式已经进行了两个多小时了，因此，我会尽量缩短我的讲话。就是说，根据21世纪的召唤，少说空话，立即行动。而且，大家可以在www. phillipines. betterfuture. com这个网站上下载我的完整的讲话。大家明白我的意思吗？（笑声，鼓掌）即使面临困难和灾难，还是要将热情好客的精神传递给每一个人。

但少说多做到底意味着什么呢？意味着我们必须使人民强大。而这也意味着我们的国家的人民也能得到好的教育，享有健康和富足的生活。还意味着为了共同的利益作出牺牲，尤其是在环境保护的问题上。因为环境并非只属于我们，也属于未来的人们。

那么对于我们来说，如何才能称得上是好邻居呢？这就需要我们具有勤勉、分享和给予的精神，不仅仅是为了各自的国家，还为了伟大的亚太大陆和整个世界。

谢谢大家！

加强政党对话，促进亚洲和平稳定和繁荣

印度尼西亚前总统　梅加瓦蒂·苏加诺

尊敬的柬埔寨总理洪森阁下，

亚洲政党国际会议联合主席，

亚洲政党国际会议各代表团，

尊敬的女士们、先生们：

能来到金边和各位领导见面，特别是能够见到亚洲政党的代表们，我感到很荣幸。我想借此机会祝贺柬埔寨人民党，它的热情招待和细致安排使我们能在这个吉祥的时刻聚集在一起。众所周知，柬埔寨和印度尼西亚一直是友好邻邦。作为柬埔寨的朋友，首先，我想对发生在上星期的金边送水节踩踏事件中的遇难者和他们的家庭传达印尼人民的哀悼之情。

各位阁下，女士们，先生们，我们的政党——印尼民主斗争党（PDIP），是我们开始民主化进程12年以来的印尼首个政党。

我们国家的五个主要指导方针包括：对上帝的信仰、仁爱、团结、民主和公平。

本着以苏加诺在1945年6月1日的演讲为基础的第一个宣言和建国五原则精神，印度尼西亚是各种宗教和民主共存共荣并携手改善社会公正和消除贫困的生动证明。亚洲政党国际会议已经成为促进亚洲不同国家政党间相互理解的积极力量。我们相信，亚洲政党国际会议期间不同政党间的对话会促进亚洲地区的民主、和平、稳定和繁荣。在这个光荣的时刻，作为印尼民主斗争党的主席，我们希望能成为亚洲政党国际会议的一员。

能加入亚洲政党国际会议，并为促进亚洲地区各民族和政党间的理解与合作作出贡献，这对我们是一个很大的挑战。

谢谢！

深化与加强亚洲政党合作
迎接亚洲美好的未来

中国共产党中央政治局委员、
中央书记处书记、中央组织部部长　李源潮

尊敬的洪森总理，
尊敬的联合主席何塞·德贝内西亚、郑义溶阁下，
各位女士、先生、朋友们：

非常高兴能在美丽的金边出席第六届亚洲政党国际会议。我谨代表中国共产党对本次大会的召开表示热烈的祝贺，向主办大会的柬埔寨主人尤其是洪森总理表示诚挚的敬意，向与会的各国代表致以良好的祝愿。

亚洲政党国际会议成立十年来，坚持以包容促合作、以务实促发展、以创新促活力，为推动亚洲各国国家关系的发展，维护地区和平稳定发挥了重要作用，成为亚洲各国政党平等交流、坦诚对话、尊重共识、密切合作的重要的多边政党论坛。我们为亚洲政党国际会议的发展壮大感到高兴和鼓舞。

新中国成立以后，特别是改革开放三十多年来，中国共产党带领中国各族人民为改变极贫极弱的面貌，追赶世界现代化步伐而不懈地努力，取得了公认的成就，但是我要说，中国现在仍然是一个发展中国家，在去年人均GDP只有3800美元，排在世界大概100位左右，中国还有1.5亿人生活在联合国设定的贫困线以下，所以中国要实现现代化还有很长的道路要走。

2010年10月中国共产党召开了十七届五中全会，审议了中国的下一个五年计划，中国将以科学发展为主题，以加快转变经济发展方式为主线，深化改革开放，扩大国内需求，提高自主创新能力，加大节能减排、保护生态环境的力度，更加注重保障和改善民生，又好又快地推动经济社会发展。

我很高兴地在这次会议上看到了我们中国的这些新的发展理念和刚才洪森总理所提到的扩大内需、应对气候变化、节约能源等发展理念是一致的，中国坚定不移地走和平发展的道路。中国是亚洲的一员，中国自古以来就有"亲人在邻，从性休牧"的传统，坚持和平发展，建设和谐社会是中国共产党和中国人民一致的主张和追求，无论现在还是将来，中国都将奉行"以邻为善、与邻为伴"的外交方针和睦邻、安邻、富邻的外交政策，永远和亚洲国家做好邻居、好朋友、好伙伴。

进入新世纪的第二个十年，亚洲各国都面临着新的机遇和挑战，进一步深化亚洲合作是亚洲各国政党肩负的共同责任和重要使命，历史告诉我们，亚洲各国和则共赢、争则俱损，希望亚洲政党国际会议坚持开放包容、深化政治悟性，坚持共同发展、促进互利共赢，坚持务实创新、拓展合作领域，为建设一个繁荣的、稳定的、和谐的亚洲发挥更大的作用。

中国共产党高度重视和加强亚洲政党的合作，始终积极地参与亚洲政党国际会议。2011年是中国共产党成立90周年，我们愿意在适当的时候和各国政党就共同关心的问题展开专题讨论，我相信在各国政党的共同努力下，亚洲政党国际会议一定能够取得更大的发展，为建设亚洲美好的明天作出更大的贡献。

祝愿大家取得圆满成功！

谢谢！

贺　信

哈萨克斯坦共和国总统　努尔苏丹·纳扎尔巴耶夫

在此，向第六届亚洲政党国际会议所有与会人员及来宾致以问候。

你们的行动最大程度地实现了亚洲各国增强友谊、合作、信任与理解的这一目标，这对共同应对全球及区域安全威胁与挑战也大有裨益。我们相信通过更广泛的交流与建设性对话，亚洲可以实现真正的安全和发展。

哈萨克斯坦从其独立伊始，就不断为增强区域稳定和安全做出努力。哈萨克斯坦主动放弃了核武器，在中亚地区建立无核区，这为创造一个更加安全的未来树立了很好的榜样。

在亚洲互动与建立互信措施会议的框架下，哈萨克斯坦积极地开展工作，以推进在军事、政治、经济、生态以及人文方面的合作。

亚洲互动与建立互信措施会议取得的成就，为这个组织在以后能够成为一个成熟的亚洲安全与合作组织奠定了基础。以和平安全与共同繁荣为目标，哈萨克斯坦发起创立了亚洲互动与建立互信措施会议。

2010年哈萨克斯坦以欧洲安全与合作组织主席的身份开展工作。欧洲安全与合作组织峰会计划于2010年12月1—2日在阿斯塔纳举行，与亚洲政党国际会议全体大会同期举行。我们相信这次论坛将会使东西方关系进入一个新的阶段，也会增强欧亚地区的互信与合作。

21世纪被喻为亚洲的世纪，因为亚洲各国在世界范围内的影响正在逐步增强。现在亚洲已经成为全球经济发展的推动力，全球发展的后危机格局已经在这个地区形成。

我相信通过我们的信任与合作，这一地区的各种难题都会成功解决。像亚洲政党国际会议这一令人尊敬的组织，在这一领域开展的工作尤为重要。

祝愿所有的与会者取得成功，成果丰硕！

可持续发展是亚洲光明未来的坚实基础

越南共产党中央委员、中央外交委员会主席　黄平军

主席先生，
尊敬的各位来宾，
女士们，先生们：

能够与其他政党一同来到美丽而又热情的柬埔寨王国首都金边市参加第六届亚洲政党国际会议，我感到非常荣幸。柬埔寨王国是越南民族传统而友好的邻邦，借此机会，我向柬埔寨民族在各个领域为建设一个和平、独立、民主和繁荣的柬埔寨所取得的成绩表示热烈的祝贺。我们正在见证一个稳定、发展与更加繁荣的柬埔寨。柬埔寨人民党与奉辛比克党为第六届亚洲政党国际会议做出的精心准备以及热情好客，给我们留下了深刻的印象。为此，我们坚信此次大会一定能取得圆满成功。

尊敬的各位来宾，我们都知道，这次大会是在一个非常重要的历史时刻召开的，因为亚洲政党国际会议已经经过了10年的发展，并形成了自己的特点。其中，那些生动具体的数字就能表明其发展成果。参加第一次大会的有46个政党，来自26个亚洲国家，而这次大会共有来自36个国家的89个政党参加，此外来自52个亚洲国家的326个政党有资格参加亚洲政党国际会议的活动。这就证明了区域政党之间，议会之间与国家之间的关系已经促进了我们向更深层次更广领域的合作，不论政党大小，不论持何种政治倾向。这次会议是亚洲政党国际会议十年来不断紧密地合作和广泛影响的证明。

尊敬的各位来宾，在此我将对我方观点作如下陈述，与大家分享：

根据我们以往的经验，我们要做的第一件事，就是务必强调可持续发展。这次会议的主题是"亚洲：追求更美好的明天"，亚洲需要未来，也有能力迎接一个光明的未来。亚洲是占地面积最大，人口最多的大洲，自然资源与人力资源丰富。亚洲经济持续快速增长，已经是世界上最有活力的地区。亚洲正在快速地发展壮大。这些都是亚洲光明未来的坚实基础。但是，亚洲现在也面临着许多紧迫的问题，包括贫困、自然灾害、流行病、食品安全、气候变化……

因此，我认为，为了亚洲美好的明天，就可持续发展交换意见与承担责任是至关重要的。这是每一个国家共同关注的事情，也是每一个地区关注的事情。了解不同国家可持续发展的意见及解决方案具有重要意义。

可持续发展是以下三个因素的和谐发展：

经济可持续发展，包括稳定宏观经济，保证经济安全，转变经济结构与增长模式，维护食品安全与能源安全以及金融机构的安全高效运作。

可持续发展就意味着经济增长应该与贯彻社会进步平等和不断改善人民生活水平联系起来。这就要求我们必须关注扶贫，缩小生活质量的差距。我们党与国家在消除饥饿与扶贫方面的经验与成功已经得到了国际社会的认可，我们愿意与其他政党与国家一起分享我们在这方面的经验。

最后，可持续发展应该特别关注环境保护，经济发展必须与环境保护及改善紧密地联系在一起；做好环境灾害预警；应用节约物质与能源的技术。这是每个国家关注的事情，也是每个国家对其他国家和整个区域的可持续发展应该承担的责任。

从这个意义上讲，就可持续发展交换意见，加强合作是个非常重要的因素。因此，我们建议亚洲政党国际会议建立定期交流机制，例如，就可持续发展开展研讨会，以提高亚洲各政党的意识以及完全掌握可持续发展在投资、贸易、开发以及能源消耗方面的观点。

尊敬的各位来宾，亚洲正在经历快速发展，但是也有战略上的部署以及利益上的竞争；亚洲的未来需要以稳定与和平为基础。在这种环境及要求下，我们建议亚洲就地区安全建立以及增强互换机制，亚洲政党在这方面的作用应该得到加强。

亚洲需要未来，也会迎来美好的未来，因此为了美好的明天应该发掘其所有潜力。亚洲是世界上人口最密集的地区，我相信大家会认同我们的观点——"人就是力量。"因此，人与人之间交流应该通过各个政党的思想和行动得到强调和表现。在这种环境下，我们尤其要谈一谈青年的作用。在任何国家，青年都是主力军，对一个国家的未来起着决定性的作用。因此，青年对亚洲的未来起着不可缺少的作用。我们提议各个政党应该考虑建立区域青年互换机制，以此为基础，进一步考虑建立区域青年组织。首先，我们建议出台一个青年政治家互换形式，以增强相互理解，分享和学习。关于如何组织，这是未来各个政党之间紧密合作的重要基础。

女士们，先生们，在此我想向大家简要地介绍一下我们的国家。经过25年的振兴，我们国家的面貌已经焕然一新，经济社会稳定；人民生活水平显著提高，多年来经济保持较快增长。我们取得的成果不仅满足越南人民的各方面的期望值，同时也履行了我们对联合国千年发展目标的承诺。

第十一届越南共产党全国代表大会将在2011年1月15日左右召开。此次大会是一个具有重要意义的政治里程碑，将对我们党的政治纲领进行回顾、修订和发展，并制定未来十年的社会经济发展战略规划。我们的目标是到2020年越南能够成为一个现代化的工业国家，到21世纪中叶成为一个繁荣、强盛、民主、平等和先进的现代化工业国家。

我们始终贯彻独立自主、自治、和平、合作和发展的外交政策；国际关系多边化及多样化，带着"越南愿意成为国际社会中的一位朋友、一位可靠的伙伴和一位尽职尽责的成员"的精神，高瞻远瞩，积极地投入到国际一体化进程中。

再次重申，我们强烈坚持越南共产党继续与世界及各地区的各个政党加强合作和友谊这一愿望。我们非常高兴也非常感谢各位参加我们党第十一届全国代表大会的国际朋友发来贺信。

作为亚洲政党国际会议常委会成员，越南共产党深深地认识到了自己在推动实现亚洲政党国际会议共同目标上的责任，而且愿意与大家分享共同的机遇和挑战。我们期望在地区政党之间进一步加强理解，增强信心与合作，共同维护亚洲的和平、稳定及繁荣发展。我还想告诉大家的是，我们党希望以后能有机会在越南举办亚洲政党国际会议常委会会议。

最后，祝愿第六届亚洲政党国际会议圆满成功。

谢谢大家！

我们有责任给人民带来稳定与和平

巴基斯坦总统特使　穆罕默德·阿扎姆·汗·斯瓦蒂

尊敬的来宾：

首先我想对尊敬的国会议员们、政治领袖们以及与会来宾们在百忙之中抽出时间来参加此次大会表示诚挚的感谢。祝贺ICAPP，祝贺它用它富于远见的领导力吸引各国富有竞争力的政府、反对党和独立政党来参加这次盛大的会议。

女士们，先生们，今天，亚洲领袖们欢聚一堂，朝着一个共同的主题——亚洲：追求更美好的明天。我们已经意识到，只有我们维持我们各国的和平，人类事业的进步才可能取得进展，而这与我们的进步、发展、繁荣和生存是分不开的。

阿富汗是通往亚洲的大门，它正面临着已长达30年的战争。无端的流血事件使它的公民处于一个战斗和防卫的环境中，而不是想着和平和安宁。现在巴基斯坦和其他一些邻国正遭受着国家的和平和安全方面的威胁。另一方面，以对抗恐怖主义为名的世界力量不仅试图占领亚洲国家的自然资源，而且试图征服其政治系统、主权以及金融和财政系统。

因此，作为一个领导人，大家都有责任协调你们的政策，给你们的人民带来稳定，带来和平，带来社会经济的安定富裕。同时，南亚国家之间也有责任解决好相互之间存在的根本问题。对我来说这个问题就是联合国长达60年悬而未决的印度和巴基斯坦之间的克什米尔问题。

如今，我们周围的世界已经发生了翻天覆地的变化，整个世界都在为人类的进步、繁荣和创新而努力。这就要求我们采用新办法来解决问题，在人力资本上投资，为社会经济结构带来正面的改变，在健康、教育、环境、扶贫上投资，还要在高等教育研究和科技上投资，从而为人类带来平等，使男性和女性得到平等的机会，使他们享受安全的、有保障的、可行的生活环境，完成他们的使命。最后，我希望这次大会的各项讨论以及各国各地区代表之间的交流能为将来形成合理的政策、框架起到深远的促进作用，使我们早日见到更美好明天的晨曦。

谢谢大家！

强化以人为本的发展理念

塞舌尔共和国外交部长　让·保罗·亚当

洪森总理阁下，

各位部长阁下，

各位与会的ICAPP领导，

女士们、先生们：

　　首先，非常感谢邀请塞舌尔共和国来参加此次大会，而且，由于很多人对塞舌尔并不熟悉，在此我首先向大家介绍一下我们这个国家。

　　塞舌尔是一个位于印度洋中部的小岛国家，而且对于亚洲来说，塞舌尔是通往非洲的第一站，连接了亚洲和非洲大陆。今天，真的很高兴能参加此次会议。作为总统特使，大家可能也注意到我作为外交部长的确有些太年轻，但是我想请大家也注意，柬埔寨总理阁下也是在非常年轻的时候步入政坛并成为政界要人的。我相信我是受到了总理阁下的鼓励。我想大家都可以看到在亚洲，年轻的力量得到充分发挥，而且我认为亚洲和非洲的未来将会延续年轻的力量，这力量将为我们的人民带来更好的发展。

　　阁下，我想首先代表塞舌尔总统阁下詹姆斯·米歇尔，就送水节期间发生的不幸事件向柬埔寨王国传达塞舌尔人民和政府的慰问。对于柬埔寨政府在事件发生后的快速反应，以及在这一困难时刻能团结全国力量，我们深受鼓舞。我代表总统，还想告诉大家，对于亚洲政党国际会议的发展，我们印象极为深刻。我们感觉到，作为非洲的一个小国家和非洲联盟的一员，亚洲有很多值得非洲学习的地方。同样，亚洲在召开亚洲政党国际会议大会上的专业经验值得非洲学习。我们向亚洲学到了很多关于共同建设的需要，以及如何运用它来在这一地区建立民主和高效的政府。我们同样感到亚洲政党国际会议在意识形态方面求同存异，这一点对于非洲来说也是极为有利的借鉴。塞舌尔认为亚洲做得非常好的一点就是以人为本的发展观念，而这对于发展来说是至关重要的。此外，我代表米歇尔总统，就柬埔寨政府提倡以人为本的发展策略向他们表达我们的祝贺。

　　亚洲和非洲的发展取决于能否将生态安全的环境保护作为可持续发展的核心任务。因此，召开的亚洲政党国际会议大会和世界生态安全大会非常适宜。我们注意到气候变化是当今人类最为关注的问题之一。而且，塞舌尔作为一个小岛发展中国家，气候变化完全影响到我们国家的发展。这两个大会的同时召开确保环境保护被列入政治议程的中心。因此，我还想对洪森总理阁下针对气候变化发表的演讲表示支持。我们相信，我们需要承担"共同但有区别的责任"，这对成功解决问题是至关重要的。因此，感谢总理阁下在这一问题上的领导力和支持力。各位阁下，女士们，先生们，我的发言可能太长了，感谢大家的关注。我代表塞舌尔政府、人民和总统，重申我们对此次盛会真诚的感谢和祝贺。

　　谢谢！

为建设生态安全和可持续发展的
地球家园而努力

柬埔寨王国政府副总理、
国际生态安全合作组织执行主席　盖博·拉斯米

尊敬的各位来宾，女士们、先生们：
大家上午好！

　　此刻，我们相聚在金边，共同出席首届世界生态安全大会。首先，让我代表柬埔寨王国皇家政府，并以东道主的名义，对出席本次大会的各位嘉宾表示热烈欢迎！同时我要感谢第六届亚洲政党国际议会和国际生态安全合组织，你们的智慧、创意和激情，让我们今天齐聚在一起！我还要感谢为本次大会的顺利召开而付出努力的柬埔寨各部门工作人员和志愿者，你们的奉献、参与、理解和支持，让我们得以在这里全身心地探讨全球生态安全发展大计。

　　二十多年前，联合国秘书长布特罗斯·布特罗斯·加利就曾指出，气候变化是一个事实，我们必须做出改变和牺牲——我们不能再靠牺牲地球和子孙后代的利益生活。那时起，我们自问：应对气候变化，应对全球生态危机，我们的工作做得够吗？显然不够！所以，我们今天才会再次会聚于此。

　　我们知道，传统意义的安全主要是指军事层面和外交层面。然而，随着国际形势的缓和，特别是进入了和平年代，人们认识到，只注意军事层面的安全已远远不够了，因为今天各个国家面临的安全不单取决于军事，它还涉及经济、能源、资源、环境、人口等等方面的要素。

　　回顾世界工业化发展历程，大多数国家实施的是"经济优先"战略，普遍经历了"先污染、后治理"的发展道路，这是人们在工业文明发展过程中付出的极为惨痛的代价。

　　目前，生态安全已经成为世界各国协调经济发展的重要因素，能源安全以及气候变化成为全球最紧迫的环境问题。世界各国正逐步把生态安全和环境保护纳入本国的发展战略，并取得了一些成果。

柬埔寨及东南亚国家自60年代以来，在经济发展方面取得了长足进展，但随之也面临环境污染的加剧、生态环境的恶化，如城市化问题、大气污染、水污染、热带雨林减少等，并且均有跨国界污染的趋势。生态环境的恶化已对东南亚国家的生存与发展构成严重威胁。

柬埔寨是东南亚最易受气候变化影响的国家之一。连年不断的水患，对柬埔寨经济社会的可持续带来了极大影响。降雨异常、旱灾、沙漠化、洪灾等天灾不但直接冲击整个国家的安危，也带来农作物、粮食及饮水安全等问题。面对生态失衡和气候变化对国民经济的不利影响，我们已经认识到问题存在的严重性，并将增强国民的环境意识、加强环境保护、处理好保护环境与发展经济之间的关系、加强生态安全领域的国际合作，列为各级政府刻不容缓的大事。

今天，我们已经认识到，人类不当剥削环境、工业国家产生大量温室气体，这些都是导致生态环境恶化的主要因素。保护生态环境才能保护我们的生存权和发展权。应对气候变化，维护生态安全，保护生态环境，应对突发性自然灾害和生态危机，需要举全人类之力！

在这里，我希望本次会议能成为促进世界各国共同致力于解决全球生态危机的一个契机，并希望大会的主题会议能够提出促进生态安全与可持续发展的具体方案。

女士们，先生们，朋友们！空气、阳光与水露孕育了生机勃勃的地球家园，也是人类赖以发展的基础。让我们满怀使命感，共同承担责任，为建设一个生态安全和可持续发展的地球家园而努力！

最后，预祝首届世界生态安全大会圆满成功！

贺　信*

马尔代夫共和国总统　穆罕默德·纳希德

2010年，世界接连经历了一个又一个与气候变化相关的灾难，如巴基斯坦的洪水，俄罗斯大火和美国严重的暴风雨。正是在这样一个年份，亚洲政党国际会议各成员相聚柬埔寨。

据预测2010年是有史以来最热的年份——明年很可能更热。即便如此，世界各国仍然无法就气候变化问题达成一致协议。

我认为我们不仅要视气候变化为挑战，更要将其看做机遇。我们不应把降低碳排放看做减少就业机会和阻碍经济发展的负担，相反，应将世界迈向绿色看做自工业革命以来最大的经济机遇。

这是一次改善提高的机会，一个让经济沿着可持续发展道路前进的机会，一个创造新财富、增加就业率的机会。

在坎昆气候变化谈判以及将于2011在南非举行的气候谈判上已取得和将要取得的进展不应被看做经济发展的障碍，而应是动力。其间所达成的协议也不应被视为发展的阻力，而应看做促进发展的方式。

马尔代夫旨在到2020年实现"碳平衡"。致力于此目标并非因为我们是激进的环境保护主义者。事实上，在一个99%的面积都是海水的国家，是不会有什么树木的。我们之所以承诺实现"碳平衡"是因为放弃使用石化燃料符合我们的经济和能源安全利益。

在马尔代夫，用石化燃料发电是非常昂贵的。而且，我们在面对变幻莫测的国外油价时的束手无策经常使我们的经济处于危险中。为了求得发展，马尔代夫需要使用可再生能源。我想对于许多在座嘉宾的国家来说，情况同样如此。

我相信，ICAPP可以在保证气候安全，以及帮助亚洲国家在可再生能源使用方面保持领先上发挥巨大作用。此外，对于ICAPP在召集亚洲政党会议方面所做的努力，我表示欣赏。

只有齐心协力才能解决我们共同面临的威胁，利用新的机遇。ICAPP这样的

---

* 本贺信由马尔代夫共和国总统特使艾哈迈德·拉蒂夫宣读。

组织在团结不同国家以及帮助这些国家增进了解，应对21世纪相互间出现的挑战发挥了关键性的作用。

祝大会取得圆满成功！

不断创新，推动全球生态安全发展

塞舌尔共和国总统詹姆斯•米歇尔贺信与
塞舌尔共和国外长让•保罗•亚当致辞

各位阁下，女士们，先生们：

能有机会代表塞舌尔总统詹姆斯•米歇尔在此发表演讲，我感到很荣幸。

由于詹姆斯•米歇尔总统出席墨西哥坎昆联合国气候会议，而无法出席此次世界生态安全大会，于是请我转达他对柬埔寨政府和人民的感激之情，感谢你们举办首届世界生态安全大会，并使生态安全这个概念得到全球性的关注。

也请允许我表达对东道主的感谢与欣赏，你们是如此的热情与好客。

同时，借此机会，代表塞舌尔人民，对上周送水节中由于踩踏事件造成的人员伤亡向柬埔寨政府和人民表达沉痛哀悼。我们看到，在这个无比悲痛的时刻，柬埔寨人民同舟共济。在这个哀悼与忧伤的时刻，我们也愿意与你们齐心协力共渡难关。

各位阁下，女士们，先生们，作为一个小岛屿国家并且是发展中国家的领导人，米歇尔总统一直在努力使国际社会在环保及生态安全的重要性方面能有更多理解与合作。

现在我将与您分享米歇尔总统代表塞舌尔人民对此次峰会所作的演讲词：

各位阁下，

女士们，

先生们，

此刻我们步入了全球管理的十字路口。

在我们面前，气候变化的威胁在不断增长，而对发展的追求也日益艰难。我们应对这些挑战的现行方法通常既速度缓慢又效果不佳。

世界生态安全大会为解决这些问题提供了一个新的务实的途径。生态安全观强调了这个事实：世界安全与我们所追求的真正的可持续发展的努力密不可分。

　　在塞舌尔这样的岛屿上，安全与生态保护的联系是显而易见的。我们的旅游业和渔业依赖于我们自然环境的可持续性。只有保护好我们赖以生存的环境，我们的财富才能得以保障。

　　这就是为什么塞舌尔在不断扩大自然保护区面积。今年前半年，我们将卢埃特岛列为新的自然保护区。这使得塞舌尔自然保护区的总面积达到国土面积的47%，这一比例也是世界最高的。

　　在我们评价坎昆气候变化会议的讨论结果时，我们也同样关注世界生态安全大会。我们希望，在气候变化方面，我们能为促进发达国家与发展中国家"共同但有区别的责任"的概念发挥作用。

　　以小岛屿国家为例，它们是气候变化最大的受害者，却不是造成气候变化的主要原因。小岛屿国家在国际发展结构中处于边缘地位。由于较高的人均GDP，我们得到发展基金的机会很少。许多小岛屿国家还债台高筑，由于不依赖高昂的商业信贷，它们很难实现发展的优先目标。

　　在我们努力适应气候变化及其相关代价时，我们经济的可持续性被进一步削弱了。

　　在很大程度上，生态安全就是找到保护地球最后乐园的正确模式。城市可以重建。但是没有什么能替代自然与我们分享的奇迹：海洋、沙滩、珊瑚礁，还有居住在那里的丰富的物种。

　　所以当发达国家正担忧他们困难重重的经济时，生态环境才是我们可持续发展的真正挑战。如果我们不首先拯救环境，最终也没有经济可以挽救。这既适用于世界经济，也适用于小岛屿国家。唯一的不同是，岛民将此视为日常生活的一部分，而世界上大部分人口脱离了养育他们的环境。

　　生态安全的部分含义是关于我们对环境的理解。我们需要在全世界范围内重建与生态环境的联系。我们举办的世界生态安全大会可以为达到这一目标作出重要贡献。

　　我们努力的方向应集中于开发可再生能源，并使发展中国家能接触到这些开发技术。

　　小岛屿国家有充足潜力去发展类似于太阳能或风能这样的可再生能源，但我们对新技术的获取经常受到限制。

　　我们应不断创新来达到这些发展。在这方面，公私合作有很大的余地，我们不应该逃避可再生资源的商业潜力。如果管理得当，这可以对发展中国家提供新的发展动力。

各位阁下，女士们，先生们，世界生态安全大会是一个理想的论坛，它为我们星球的可持续发展创造了一个新的动力。我们应汇集各种想法，也许他们在理论上相互矛盾。我们需要发展，也需要保护环境。我们需要满足眼前的需求，也需要考虑将来的事情。

我们很高兴地看到这个论坛聚集起如此多聪明的头脑，你们的创新想法会解决这些不可调和的矛盾。

我要特别感谢蒋明君博士，他辛勤的工作才使这次大会得以召开。把我们汇聚在一起的动力反映了经受考验的愿望——真正将生态安全纳入全球管理的重心。

可持续发展的首要障碍与可持续发展不相协调的首要问题就是继续现状。但我有信心此次世界生态安全大会能改变这种状况，并借此改变世界。

感谢您的耐心倾听，并希望诸位都有机会参观塞舌尔，并能亲自体验印度洋的美景。

各位阁下，女士们，先生们，我很高兴能代表总统发表讲话。他的讲话重申并反映了塞舌尔对大会理念所做出的努力。

我们期待着会议的结果，我们相信这些结果能带来真正的不同。

谢谢！

贺　信*

乌干达共和国总统　约韦里•卡古塔•穆塞韦尼

尊敬的洪森总理，

尊敬的政府首脑们，

ICAPP秘书长和成员们，

各位来宾，女士们，先生们：

请允许我向你们转达乌干达共和国总统约韦里•卡古塔•穆塞韦尼阁下以及乌干达政府和人民热烈而诚挚的问候。请允许我转达乌干达政府和人民对于送水节上发生的伤亡向柬埔寨政府和人民致以衷心的问候和同情。

昨天，我和其他代表们有机会参观了波布罪恶馆。这个博物馆是你们的国家在20世纪70年代所遭受的惨剧的鲜活证据，也是我们的政府委任的本应防卫和保护其人民的独裁者阿明的罪证。乌干达和柬埔寨有着共同的动荡和血腥史。更巧的是，这段历史都在70年代，几乎发生于同一时间，结束于同一时间。乌干达见证了1971和1979年之间阿明的独裁统治。它在本应保护好人民的领导者的领导之下，却失去了很多人民。与乌干达一样，柬埔寨在如今洪森总理、柬埔寨人民党以及政府的英明领导下，终于得以浴火重生，逐渐转变成为稳定和经济快速发展的榜样。总理先生，请允许我热烈祝贺您和您的政府，祝贺你们取得这些惊人的成就。我也祝贺ICAPP第六届年会，祝贺它在增强亚洲各政党之间的合作方面取得的成果。我们欢迎其他为了将这种合作和联系扩大到非洲地区甚至世界其他部分所做的努力。我们也期待那一天的到来，正如何塞•德贝内西亚阁下所说，我们期待那个我们能够举办一次全球政党大会的时刻的到来。ICAPP的成功，的确对于正在发展中的世界来说是一个极大的鼓舞，同时也告诉我们，大家协力合作，一定能达到我们共同的目标——更加美好的生活。

最后，我想对于邀请乌干达来参加此次大会表示感谢，同时，特别感谢洪森总理以及柬埔寨政府和人民对于我们的热情招待，感谢你们在金边这个美丽的城市成功举办ICAPP第六届大会和首届世界生态安全大会。

谢谢大家！

---

* 本贺信由乌干达共和国总统特使瓦吉多索宣读。

贺　信*

俺罗斯联邦委员会主席　谢尔盖·米罗诺夫

尊敬的蒋明君先生，
亲爱的朋友们：

我代表俄罗斯联邦委员会向首届世界生态安全大会全体嘉宾致意！

你们的行动，无疑是致力现代人类极其迫切的问题——地球生态安全。大自然保障着人类生存的基本条件，但我们，为满足自身需求，有时疯狂地消耗着自然资源。

俄罗斯著名学者维尔纳斯基在自己的著作中指出，人类在地球表层变化进程中起着非常重要的"地质因素"，当今的任务在于，确保这些自然环境的变化不会危害人类本身与整个自然界。如今我们必须共同采取措施，保持生态环境平衡、恢复人类适宜生活的自然条件。

我坚信，首届世界生态安全大会将会在提高全球共同体环境保护工作效率、开发生态清洁工业科技等领域中作出重要的贡献。

祝你们工作富有成效，在专业与创新上获得成功。

---

* 本贺信由国际生态安全合作组织执行主席、柬埔寨副总理盖博·拉斯米宣读。

生态安全需要我们共同呵护

尼泊尔民主共和国总理　马德哈夫·库玛·尼帕尔

尊敬的大会主席，

各位嘉宾，女士们、先生们：

　　大家好！非常荣幸出席第六届亚洲政党国际会议和首届世界生态安全大会。本次大会的主题是"和平发展与生态安全"，我相信与会的各位代表也和我一样，都在密切关注生态安全问题，并且都希望通过自己的绵薄之力推动和平发展，共同保护我们的地球家园！

　　从国际关系来看，以环境和生态为主题的国家交往，可以追溯到20世纪70年代。1972年，由于环境问题日益突出，全球兴起了一场规模浩大的绿色运动，当年6月，联合国人类环境会议在瑞典斯德哥尔摩召开并发表了《人类环境宣言》，这成为世界"环境外交"的伊始。1992年6月，全球180多个国家的首脑聚首巴西里约热内卢召开联合国环境与发展大会，会议制定了《21世纪议程》，并签订了气候变化框架公约、生物多样性公约，发表了里约环发大会宣言和关于森林问题的原则声明，这次会议，成为世界"环境外交"的一大里程碑。2009年12月，联合国气候变化框架缔约国会议在哥本哈根召开，190多个国家的首脑出席会议。就在本次大会召开之际，联合国气候变化框架缔约国新一轮会议正在墨西哥坎昆举行。

　　从全球范围来看，生态安全已经成为一个全球关注的时代主题。我们每天都可以通过各种媒体，看到来自世界各地的自然灾害与突发性生态灾难的报道。旱灾、洪涝、台风、风暴潮、冻害、雹灾、海啸、地震、火山、滑坡、泥石流等灾害灾难产生的原因，除了自然变异就是人为影响。维护生态安全，保护生态环境，需要我们的共同努力。

　　尼泊尔联邦民主共和国位于喜马拉雅山脉中段南麓，以高山国家著称，气候宜人，风景美丽，文化特色鲜明，是著名的"珠峰胜境、佛陀圣地"。尼泊尔一向重视保护生态环境，从冻土带到大草原，从热带雨林到灌木丛，尼泊尔建设了许多野生动物保护区、国家公园和保护区域，因此旅游业是尼泊尔的一个重要经济支柱。与亚洲其他地区相比，尼泊尔已设法保护了更多种类的濒危动植物。现在，我们的国家林业与土地资源部正在实施尼泊尔生态安全与可持续发展项目，将致力于保护森林资源，关注生物多样性区域，促进湿地的保护和开发，启动社区自然灾害管理计划，加强气候变化和自然资源管理，提高城市空气质量，改善卫生健康环境，开发生态能源，建立在收益共享基础上的资源可持续发展机制。

2008年，国际生态安全合作组织曾派代表团到尼泊尔考察，并决定与尼泊尔国家林业与土地资源部合作建立"兰毗尼国际生态安全示范基地"，进一步推进国际生态安全自然保护区建设，以及生态安全与城市可持续发展项目。我们也欢迎其他国际组织和非政府组织，到尼泊尔参观、考察，支持我们建设一个生态安全与可持续发展的国家。

谢谢！

用"3E理念"推进人与地球的和谐发展

联合国副秘书长、坦桑尼亚国土住房部部长　安娜·蒂贝琼卡

尊敬的大会主席，

各位嘉宾，

女士们，先生们：

大家上午好！

我很荣幸来到美丽的柬埔寨首都金边，出席首届世界生态安全大会。感谢大会组委会的安排，让我有机会与在座的嘉宾商讨一个共同的话题，也就是全球迫在眉睫的生态安全与可持续发展问题。

在我就任联合国副秘书长和人居署主任期间，我和我们的团队提出了一个"3E理念"，（即Ecology、Economy、Equity，生态、经济、公正）。三个月之前，我从联合国人居署卸任了，但我想"3E理念"的价值，是没有阶段和时间限制的，它值得我们每一个与会者关注与思考。

就在不久前，我从一份资料中获悉，2010年，城市人口首次超过了乡村人口，乡村占主导的社会正加快向城市占主导的社会转型。毫无疑问，我们生活在一个史无前例的、快速城市化的时代，它正在改变我们生活的世界，我们需要做好准备。这其中的一个突出问题，就是气候变化。因为这个问题已经威胁到了人类的生存和发展，需要引起世界各国的广泛关注。比如坦桑尼亚的达累斯萨拉姆是个海滨城市，它地势很低，如果附近海域海平面上升1米多的话，它就有可能被淹没。所以，我们首先要应对气候变化，共同创建一个生态安全、人与自然协调发展的和谐世界。

早在2004年，我在意大利出席世界政治论坛时认识了蒋明君先生，他提出的生态安全理念感染了我。在后来联合国人居署主办的三届世界城市论坛上，蒋明君都带领他的团队参加了，并且每届论坛他们都举办了"生态安全与城市可持续发展"专题会议，并与联合国人居署建立了很好的合作伙伴关系。

当前，我们面临的重大挑战之一，是更多人追求现代化生活的强烈意愿与全球自然资源和生态环境承载能力之间的矛盾，而且这个矛盾日益尖锐。现在

的趋势是每个人都在谈生态，但他们的想法并不真正与行动一致。

从全球来看，各种突发性自然灾害与生态灾难几乎每天都在发生。当灾害灾难发生之后，我们再去追根究底，却发现问题的根源还在于我们人类自身。我们对自然的过度索取，人类无止境的欲望，打破了人与自然之间的平衡。所以，实现生态安全和可持续发展，首先取决于我们到底要选择怎样的现代化发展方向、方式与进程，政府应该更加积极地去寻找一系列新的解决方案，实施积极的社会管理和发展规划，促进社会公平。

同时，非政府组织和社会公民也要扮演积极的角色，因为政府的监管不可能面面俱到。它们要发挥自己身处某些特定社区的作用，向政府和社会提出建议和创意，借助网络和论坛的力量，促进整个社会和谐相处，共同致力于生态安全与可持续发展的推进。

最后，预祝此次大会取得圆满成功！

实施"环保项目抵债计划" 减缓气候变暖

亚洲政党国际会议创始主席　　何塞·德贝内西亚

在这个美丽的季节，我很高兴能应邀在金边参加与第六届亚洲政党国际会议联袂举办的首届世界生态安全大会。对于能够参与此次国际生态安全合作组织所做的历史性行动，以及能与柬埔寨副总理盖博·拉斯米领导下的奉辛比克党、柬埔寨总理洪森和副总理索安领导下的人民党、柬埔寨议会领导人韩桑林和谢辛合作，我深感荣幸。

环境保护——ICAPP的行动计划

环境保护当然要被列入亚洲政党国际会议的基本行动纲领中，这也是国际生态安全合作组织努力追求的最大目标。我们意识到任何一代人都不能随意地、不加节制地攫取地球资源，而要为了我们的后代好好保护地球。

我们也深切意识到生态危机已经成为我们共同家园的最可怕的威胁。我们几十年来对地球资源的挥霍以及对地球的漠视现在正在遭到报复。这些报复包括各种各样的生态灾难，诸如暴风雨（雪）、洪水、干旱、湿地面积的锐减、沙漠的扩张和物种的灭绝。

而且，像往常一样，遭受这些所谓的"自然"灾害最严重的总是那些最穷的国家。灾害在瞬间就将几十年来人们辛勤耕种劳作、省吃俭用积累的物质财富一扫而光。

培育生态观念，扭转恶化趋势

为扭转生态环境急剧恶化的趋势，人类必须学会与自然和谐相处。我们人类再也不能傲慢自大、目空一切，以所谓宇宙的主人而自居了，必须开始像上帝公正地看管人那样管理所有生灵，为地球改换新容颜。

幸运的是，人们正开始在日常生活中培育自己的生态理念和文化。有机食品已经在全世界范围内普及，在欧洲尤其受到欢迎。中国共产党已经宣布将"生态文明"作为国家发展战略。在联合国的倡议下，国际社会已经开始就缓解全球变暖和气候变化进行磋商。

关于气候变化的谈判问题

气候谈判最困难的地方在于新兴工业国家（NICs）不肯接受国际上约定的降低碳排放量的标准，原因在于这些国家想得到达成这些目标所需的经济费用和技术补偿。它们认为，从历史上来看，它们自身的平均碳排放量要远远低于工业化国家。在可预见的将来这一现象还将继续。

为最不发达国家量身定制"环保项目抵债计划"

上述问题使得去年在哥本哈根举行的气候谈判陷入僵局。为解决这一难题，菲律宾提出了详细的"环保项目抵债计划"，以此作为对"债转股计划"的补充。此计划旨在帮助100个负有高额债务的中等收入国家对抗贫困，同时也为联合国千年计划提供资金支持。

一方面，此计划属自愿执行，并不要求国际社会中的债权人减少其债务国的债务，也不要求任何富国的立法机构投入新的资金，诸如美国国会、英国议会、日本国会、德国联邦议院，中国全国人民代表大会和法国国民议会。

另一方面，此计划执行起来快捷简便。此计划允许债权国和贷款机构将所收到的最多高达50%的应还债务金额转移到有利于投资者权益的由债务国投资的环保项目中。

此计划包括多个项目，诸如植树造林、水资源保护、使用替代化石能源的清洁能源、公共住房建设、健康保健、教育、生态旅游以及其他社会基础设施建设。

具有商业可行性的社会项目

这些项目中的大部分都能带来潜在的商机，比如各种费用、使用者的税金、租金和销售收入。其中通过植树造林项目还可以获得《京都议定书》规定的碳排放额度。通过木材的一条龙销售也可获得丰厚利润。生态旅游同样也能带来收入和就业机会。

公共住房建设项目使不富裕国家的政府能够安置那些住在河畔以及易受泥石流影响的山区成千上万个贫困家庭。

作为这些项目的股权持有人，债权国有权监督这些由它们投资的项目。

而对于那些不富裕国家，这些用环保抵债的计划，特别是植树造林项目，可以在全球经济衰退的趋势下对经济产生巨大的刺激，并且创造大量的工作岗位。从债务还款中转出的资金直接投资于环保项目中，由此推动不富裕国家的经济发展，并且为技能较低的民众创造出大量工作机会。

被联合国大会认可与支持的计划

菲律宾最初提出的"债转股计划"已得到联合国大会，德国、意大利和西班牙政府以及77国集团加中国，亚洲政党国际会议，第11次东盟国家元首及总理峰会和东南亚国家联盟的认可。此计划旨在为联合国千年计划在100个负债最重的国家实施的削减贫困的项目注入资金。

中国在应对气候变化问题中的态度

在所有大国中，中国在应对气候变化的问题上将自己置于穷国的立场上。

在回复菲律宾的提议中，北京方面派出的气候谈判特别代表于庆泰说，在联合国气候变化框架公约下，富国和穷国有着"共同但有区别的责任"。

公约规定，发达国家要向发展中国家提供资金，以补偿新兴工业国家在达成减少碳排放目标时所需的资金。

中国对出资问题的声明确定了什么才是富国和穷国在应对气候变化时必须通过共同合作来解决问题。

国际生态安全合作组织所做的大胆尝试值得赞扬

总而言之，我必须得说"环保项目抵债计划"是国际社会在确保成功应对环境恶化和气候变化问题上所采取的诸多复杂措施之一。

而且我必须提醒大家的是，不要希望各国能在12月份在墨西哥城举行的联合国气候变化大会第二轮谈判上达成大规模的一致意见。

在这种情况下，这次首届世界生态安全大会是国际生态安全合作组织所做的一次大胆而可取的努力。在此，我要感谢蒋明君阁下这位洋溢着激情的领导者，还要感谢所有参会的国家领导人，感谢你们的大力支持。

既然你们已经决定了所要做的事情，我们亚洲政党国际会议很荣幸能与你们一道确保我们地球家园的生态安全。

我代表亚洲政党国际会议，国际生态安全合作组织和我们新的合作者——亚太中间党派民主国际，并且以我个人的名义，祝愿大会取得圆满成功。

敦促政治家以绿色理念治理国家

我们国际生态安全合作组织和亚洲政党国际会议的工作者能做些什么呢？

维护生态安全是一项长期事业，我们要做世界生态安全大会的政治发言人，我们可以为保护环境召开亚洲政党国际会议，我们为应对气候变化结成前线同盟。

我们要使普通民众认识到生态保护的紧急性，以及他们所拥有的力量。要让他们知道，通过他们共同投票选举，可以促使当政者以绿色理念来治理国家，确保在每一项公共政策的决策中体现出生态安全。

应将生态安全、环境保护写进宪法和党纲

通过亚洲政党国际会议、国际生态安全合作组织和亚太中间党派民主国际的努力，我们可以尽最大的努力——甚至敦促——将生态安全，即防止环境恶化以及保护环境的公共政策写进每一个国家的宪法，每一个政党的党纲以及每一个文明社会的宪章。

我相信这些是我们能够尝试和做到的最基本的事情，这也表明了我们国际生态安全合作组织和其他卓越的领导人在已经得到重视的全球环保事业上的团结一致。

我代表亚洲政党国际会议和国际生态安全合作组织，并且以我个人的名义，祝愿大会取得圆满成功。

谢谢大家，祝愿大家拥有美好的一天！

保护生态环境就是保护人类文明

蒙古国前总理　门德赛汗·恩赫赛汗

尊敬的大会主席，
各位嘉宾，
女士们、先生们：

大家好！非常高兴能够来到柬埔寨金边参加首届世界生态安全大会。当今世界，生态安全已成为一个家喻户晓的热门话题。我很荣幸能借助这次大会的平台发表我的看法。在此，我谨代表蒙古政府对首届世界生态安全大会的召开表示热烈的祝贺！

人类的生存与发展依赖于自然，同时也影响着自然的结构、功能和演化过程。人与自然的关系体现在两个方面，一是人类对自然的影响与作用，包括从自然界索取资源与空间，享受生态系统提供的服务功能，向环境排放废弃物；二是自然对人类的影响与反作用，包括资源环境对人类生存发展的制约，自然灾害、环境污染与生态退化对人类的负面影响。随着人类社会生产力的不断发展，人类开发利用自然的能力不断提高。人类社会发展史其实就是一部生态环境的血泪史。自然资源枯竭，全球气候变暖，臭氧层空洞扩大，大气、水、土壤污染严重，物种灭绝加速，土地荒漠化严重，洪灾泛滥……地球正用它的方式惩罚人类对自然近乎疯狂的掠夺行为。

蒙古国位于亚洲中部，是一个有着独特民族风情与自然地理景观的国家，同时也是一个资源丰富的国家。现在已经探明的矿产就有铜、金、银、铀、铅、锌、稀土、铁、萤石等八十多种。

可是，随着自然资源的过度开发利用，目前蒙古国也正面临着环境资源与环境产权、空气污染、草原畜牧业资源的利用、森林的保护与利用，以及水资源的布局与利用等多种环境压力。其中最为严重的问题，便是荒漠化问题。蒙古国已被联合国列为受荒漠化威胁最严重的11个国家之一。截至目前，蒙古国已有72%以上的土地遭受了不同程度的荒漠化。

森林安全对防治沙漠化和全球变暖起主要作用。保护森林唯一的也是最有效的方式就是授权给居住在森林地区的地方居民。因此，蒙古政府在2007年对现行的《森林法》进行了一项重要的修订：通过合同，居住在森林旁的居民可以拥有一定的森林基金。由于新的修正法案出台，许多森林用户组得以建立，他们在保护和可持续利用森林资源方面起到了重要作用，在很大程度上减少了实行森林基金地区的非法伐木和森林火灾现象。这同时也是对保护自然和改善居民生计的重要贡献。

事实上，荒漠化不仅仅是蒙古国的生态难题，也是世界的生态难题。全球约三分之一的人口生活在荒漠化危害地区。荒漠化就像瘟疫一样，也同样困扰着北美、澳大利亚、中亚以及中东地区。我始终认为人类无论怎样推进自己的文明，都无法摆脱文明对自然的信赖和自然对文明的约束。自然环境的衰落，也必将是人类文明的衰落。所以，让我们一起为全球生态环境的改善作出贡献吧。

谢谢！

发挥国际组织在生态安全中的作用

上海合作组织首任秘书长、
国际生态安全合作组织联合主席 张德广

尊敬的大会主席，
各位嘉宾，
女士们、先生们：

大家好！今天，我们相聚在柬埔寨金边，共同出席第六届亚洲政党国际会议和首届世界生态安全大会。借此机会，我谨代表国际生态安全合作组织，对出席本次会议的全体嘉宾表示最热烈的欢迎！

当今世界处在一个经济全球化深入发展、政治多极化加速前行的时代，这个时代的显著特点，就是没有哪一种力量能够单独主宰世界事务，特别是一些关系到全球发展的问题，如应对气候变化，维护生态安全，保护生态环境，推进城市发展，减少贫困，实施节能减排等，都需要国际间的合作与交流。

人类在21世纪面临的最大问题，是如何共同应对各种复杂的挑战，推进全球可持续发展。正因如此，气候变化与生态安全已成为当今国际社会的热门话题，"环境外交"、"生态外交"已成为世界各国外交中的新课题。

本次大会的一大特点，是亚洲政党国际会议成为生态安全大会主办方之一，柬埔寨人民党、奉辛比克党发挥东道主的作用，国际生态安全合作组织、联合国大学、中国国际问题研究基金会参与主办和协办本次大会，体现了非政府组织在顺应全球化与世界多极化发展趋势、推动全球生态安全与可持续发展中的独特作用。

近年来，许多国际组织、非政府组织，在生态安全和可持续发展领域做了大量的工作，它们通过开展各种层次的对话与合作，成为调停和解决气候变化与生态安全领域争端的重要渠道。

蒋明君博士是国际生态安全合作组织的总干事。他为推动世界生态安全事业作出了重要贡献，这至少涉及以下三个方面：

一、2006年，他创建了国际生态安全合作组织，而在当时，生态安全在世界上还只是个鲜为人知的概念。创建国际生态安全合作组织，体现了他在这一领域所具有的远见卓识。

二、经过多年孜孜不倦的研究并依据其丰富的阅历和经验，他提出了一套系统的生态安全理论，近日出版的《生态安全学导论》一书，集中体现了他的研究成果，从而为这一领域深入的科学研究开了先河。

三、在深刻认知世界发展趋势的基础上，为维护生态安全，促进生态文明，打造国际合作的有效平台，他克服重重困难，成功地在柬埔寨举办了首届世界生态安全大会，与会的有来自数十个国家的党政官员、非政府组织和企业、艺术机构的领导人和专家，从而开创了生态安全国际合作的新模式。

维护生态安全是每个国家、每个社会阶层、各行各业的人士，每个地球公民的共同责任，希望国际社会能有更多人士像蒋明君先生那样，热心致力于生态安全事业。相信本次大会的举行，将能够建立一个广阔的有效的大平台，吸引更多国家政府、政党议会、国际组织、非政府组织、金融机构、企业集团等参与进来，共同推进全球生态文明，应对气候变化，维护生态安全，保护生态环境，为实现联合国千年发展目标发挥积极作用。

谢谢！

二、主旨演讲

（气候变化与生态安全）

积极应对气候变化危机的挑战

柬埔寨王国政府环境部部长　莫马烈

主席阁下，

蒋明君总干事阁下，

来自国内外尊贵的客人们：

今天，我很高兴，并且很荣幸代表柬埔寨皇家政府在首届世界生态安全大会上作主旨演讲。在此，我对各位嘉宾的到来致以热情的问候，并欢迎各位来到柬埔寨首都金边。我向各位参会嘉宾表示衷心的感谢，同时我也希望，你们能充分享受到柬埔寨的迷人风光，尤其是承载着世界文化遗产吴哥窟的光辉之地——暹粒。

各位朋友、女士们、先生们，10年前，柬埔寨才从长达60年的内战阴影中走出。历尽冲突之后，在恢复和发展经济、基础设施和社会服务的进程中，我们面临着许多亟须解决的问题和挑战，如改善民生，消除贫困并确保可持续发展，而这些都将遵循皇家政府矩形战略的第二阶段目标和2009—2013年国家战略发展修订计划。为了支持作为国家经济骨干和扶贫关键的农业部门发展，柬埔寨王国皇家政府在发展基础设施和开发灌溉系统上已经取得了重大进展。目前我们已经设立了一个宏伟目标，即到2015年使柬埔寨成为出口量达100万吨的大米净出口国。同时，我们也在进行渔业改革，以确保鱼类继续成为全国人民获取蛋白质的重要营养来源。

另外，全面的生态与环境法律框架已经启动，以支持我国对自然资源的合理利用并确保高质量的生态环境。在覆盖全国总土地面积约25%的保护区和其他森林保护区内，我们的保护行动已初见成效。我们有一个雄心勃勃的目标，到2015年，我国的森林面积将增至全国土地总面积的60%。目前，我们正致力于使长达435公里的沿海地带获得"世界最美海湾俱乐部"的提名。让我们颇为自豪的是，在"平衡发展与保护，以保护支持发展"的方针指导下，通过这些年的努力，贫困范围已从1990年的50%减少至27%，同时，我们成功保持了较高的环境质量。这也使发展对自然资源的压力相对减少。

　　总的来说，柬埔寨目前社会和政治稳定、社会和谐、经济稳步增长，这要归功于柬埔寨王国总理及国家气候变化委员会名誉主席洪森阁下的卓越领导。我们在地区和国际舞台上扮演的角色也愈加重要。柬埔寨在应对区域和全球性挑战、促进合作等方面提出了多项关键的区域性措施。目前，我们已经向众多国家派遣维和部队，以体现我们的人道主义任务。

　　我们正处于这样一个时代，气候变化已成为影响国家、地区和全球的一个关键因素。作为大湄公河次区域的一个农业经济国家，柬埔寨受到了气候变化的严重影响，越来越多的科学证据显示，湄公河将受到气候变化的严重影响。气候变化对水域、渔业、湿地生态功能的影响会对人们的生活产生严重的消极作用，并使来之不易的发展成果付诸东流。

　　去年，柬埔寨遭遇了强台风凯萨娜，大量人员伤亡，农作物、基础设施和房屋也遭到严重破坏。今年，在本就姗姗来迟的雨季之后，又迎来了连绵不断的极端降雨。就在一个月前，在金边附近的城市塔克莫，一场降雨量高达50厘米的暴雨倾盆而至，导致大面积山洪暴发。

　　柬埔寨王国政府在洪森总理阁下的领导下，对这一新的挑战有着清醒的认识。而作为联合国气候变化框架公约及《京都议定书》的缔约国，柬埔寨一直在尽最大的努力，履行其在这些国际条约下的承诺。最近，柬埔寨参加了第15届气候变化框架公约缔约方大会。在大会上，柬埔寨首先立场明确地表达了将减少温室气体排放的承诺；其次，为适应气候变化，弱国有着大量资金和技术支持的需求；最后，柬埔寨发出了对发展中国家转让技术的呼吁。此外，我们强调发展中国家根据气候变化做出调整的重要性和紧迫性，呼吁发达国家履行气候变化框架公约下的承诺，大幅削减其排放量，并有针对性地向发展中国家提供财政支持。以上是我们连同77国集团和中国一贯坚持的立场。

　　我们深信，为了完成联合国气候变化框架公约的最终目标，大气中温室气体的浓度不能超过百万分之四百五十，这样的话温度上升就不会超过两度。因此，各国应本着联合国气候变化框架公约中的"共同但有区别的责任"的原则，积极推行减排。就这一点而言，一些发展中国家在温室气体减排上取得的成就让我们深受鼓舞，其中就包括中国。目前，中国正在成为可再生能源开发和节约能源的主要国家。

　　从区域层面上讲，根据1995年《湄公河协议》，我们已经在积极推动与东盟、大湄公河次区域和湄公河委员会在区域环境与气候变化上的合作，这个协议主要关注湄公河流域水资源及其他资源的可持续发展、利用、管理和保护这些方面的合作。关于湄公河支流和湄公河上下游沿岸的所有开发项目对社会与环境的影响及其成本效益，我们希望进行认真全面的评估。我们也呼吁进行合作、对话和信息共享以促进公正、平衡与可持续地利用湄公河重要的共用资源。从这一点来讲，柬埔寨大力支持流域发展计划（BDP）的完善与贯彻，促进、支持并通过合作以协调该计划最大程度地发挥出使所有湄公河沿岸国家获得可持续利益的潜力，同时防止湄公河流域水资源的浪费。我们大力支持水资源使用项目的持续实施，以改善流域共享水资源的使用和管理，同时提高对生态平衡重要性的理解。所有这一切都必须适应气候变化的背景，并且将适应这些变化所做出的改变融入到所有的发展政策、计划和行动中去。

朋友们、女士们、先生们，柬埔寨皇家政府在洪森总理阁下强有力的领导下，坚定地为应对上述全球挑战作出自己的贡献。以现有的国家政策法规、区域性以及国际性协议为基础，我们立志要与世界各国一同应对气候变化、扭转重要资源恶化的局面，为子孙后代保护我们的地球。从地区性讲，我们有充分的政治意愿将湄公河树立成区域合作的典范，成为和平共处与共同繁荣的标志。

我们人类总认为自己拥有超过地球母亲的能力，为了满足自己的目的、需求以及经常伴随人类的贪婪，从而任意改变地球。近期出现的生态危机再一次提醒我们，大自然的忍耐是有限度的，如果我们想要继续作为一个物种生存的话，就不能再像以前那样对待大自然。现在的我们也许已经走到了临界点，气候变化和其他生态环境的挑战开始改变人类文明发展之路。当前的形势已经非常紧迫，我真切地希望首届世界生态安全大会能促进全球共同努力，拯救地球，最终挽救人类。在这为期两天的大会中，我们大家一定会感受到团结的气氛，为保护我们赖以生存的地球而共同分担的责任、相互之间的理解、各位的慷慨奉献以及光明的前景，对此，我满怀信心。

我希望本届大会取得圆满成功，谢谢大家！

生态安全：实现和平与发展的基础

国际生态安全合作组织总干事　蒋明君

尊敬的门德赛汗·恩赫赛汗总理阁下，安娜·蒂贝琼卡副秘书长，阿瓦鲁·乔德哈瑞副秘书长，盖博·拉斯米副总理阁下，莫马烈部长，张德广部长，杨慎部长，各位来宾、女士们、朋友们：

　　首先感谢柬埔寨王国皇家政府、亚洲政党国际会议、柬埔寨人民党、奉辛比克党、联合国大学、中国国际问题研究基金会等机构对本次大会给予的关注和支持；同时也感谢各国政党领袖、国家政要、专家学者莅临本届会议。

　　下面，请允许我代表国际生态安全合作组织向大会报告工作，请审议。

一、当前的生态危机状况与对策

　　众所周知，近几年全球性自然灾害和重大生态灾难接踵而来，地震、海啸、飓风、火山喷发；洪水、干旱、泥石流、森林大火、土地荒漠化、沙尘暴、水资源污染、环境污染、工业污染、湿地锐减、物种灭绝；原油泄漏导致海洋生态破坏，不科学的水利工程与地下水资源的过度开采所形成的地质灾害；艾滋病、非典、禽流感、疯牛病、口蹄疫、霍乱等流行性传染病。这些由气候变化与人类经济活动引发的自然灾害与生态灾难已对人类生存和国家发展构成严重威胁，而且灾害发生的频率、人员伤亡以及经济损失的程度仍在不断增加。毋容置疑，一场局部战争要有一个漫长的外交过程，而一次突发性的生态灾难却是瞬间的，其造成的人员伤害和经济损失远超过一场局部战争。

　　早在1998年，我在担任亚太地区国际协调委员会秘书长期间，便打破传统生态学的理论束缚，首次提出："21世纪最大的政治问题，一是生态安全，二是资源安全"。因为长期以来，国家与国家、地区与地区之间所发生的纠纷与冲突都与这两个方面有直接关系。例如，苏丹达尔富尔地区、印度与巴基斯坦以及孟加拉国的水资源纷争足以说明这一点。我的这一理念提出后，很快引起了俄罗斯、美国、联合国、欧盟等一些国家和国际组织的重视。1999年初，俄罗斯成立了由俄罗斯国家安全委员会领导的国家生态安全委员会；同年末，美国也成立了国家生态安全管理机构。近年来，美国在生态安全研究方面虽然关注本国问题，但重点是全球问题。他们认为，世界范围内的生态危机已威胁到美国本土的繁荣，但他们也认识到，生态危机造成的地区冲突或国家冲突，都可能使美国卷入代价昂贵甚至危险的军事干预。

二、第一个五年回顾与第二个五年计划

1．第一个五年回顾

2006年2月，面对频发的生态危机，由中国策划，并在联合国机构的支持和参与下，在香港创建了一个致力于通过与各国政党、议会、政府机构、非政府组织、企业集团合作，来维护生态安全、解决生态危机，以实现联合国千年发展目标为己任的国际生态安全合作组织。该组织创建以来，围绕国际会议、项目合作、科学研究三个方面，开展了富有成效的工作：一是在荷兰海牙举办了首届城市外交会议；在布鲁塞尔与欧盟议会举办了世界和平用水会议；分别在西班牙巴塞罗那，加拿大温哥华，中国南京，巴西里约热内卢参加了第二、三、四、五届由联合国人居署和成员国政府主办的世界城市论坛。期间我们主办了生态安全与城市可持续发展专题会议和展览会；还在中国北京与联合国大学、中国国际问题研究基金会、欧洲PA基金会共同举办了食品安全与生态安全论坛；五年来我们共主办和参与主办了37次重要的国际会议，并签署《世界城市外交宣言》、《世界和平用水公约》、《世界生态安全宣言》等战略性文件六份。二是依据生态安全管理体系指导建设了三个国际生态安全示范城市，四个新农村建设的最佳范例，五个国际生态旅游示范区。上述项目的实践，体现了经济、生态、社会的平衡发展。三是积极推广生态安全技术。由邓楚柏研究员利用工业污水、垃圾渗透液、生活污水，添加特殊微生物群落发酵合成的微生物生态能源——海力富项目，填补了世界生态能源领域一项空白，受到联合国、欧盟等组织的重视和关注；由史汉祥研究员主持研究的太极炉渣脱硫+盐碱地治理技术，用于治理盐碱沙荒地，既还蓝天，又造绿地，是以废治废、实现资源综合利用的最佳范例；由朱启江研究员主持研究的重度盐碱地治理技术，可使大量重度盐碱地变成良田；由林占熺院士主持研发的菌草种植技术，在南非、莱索托、卢旺达等国家开展技术扶贫，既解决了当地民众就业，又消除了贫困，受到所在国家政府的关注和支持；由王厚德院士主持研究的植物乳酸菌技术，用于生态农业开发，在生态种植、生态养殖方面发挥了重大作用；由彭培根院士主持研究的建筑隔震体系是目前较为安全、适用、经济的抗震技术之一，已被联合国采纳；由王嘉猷院士主持研究的"酸膜"种植技术，对于防沙治沙和荒漠化治理起到良好的作用，受到中国领导人及有关部门的关注和

重视。

五年来，我们还与中国环保部、国家林业局共同主办了首届生态环保万里行活动，此次活动行程约2.2万公里，环绕了近半个中国，对沿途各省市的空气污染，水资源污染以及废水、废气、废油排放情况进行调研，及时向当地政府机构提供咨询，并协助制定改进措施。我们还积极参与5·12汶川特大地震和海地特大地震的救援和灾后重建工作。先后出版《国际生态与安全》杂志27期，出版国际生态安全系列丛书9部。

近年来，我们还提供资金援助中国、坦桑尼亚、尼泊尔等发展中国家、欠发达国家和贫困地区的教育、扶贫、课题研究。目前，我们已与83个国家和国际组织结为战略合作机构，并如期完成第一个五年的各项工作，为第二个五年奠定了良好的基础。

2. 第二个五年计划

我们将以第六届亚洲政党国际会议和首届世界生态安全大会为契机，以《亚洲政党国际会议（金边）宣言》和《世界生态安全大会（吴哥）议定书》为准则，深入谋划第二个五年发展战略，并制定第二个五年计划：

一是抓好三个主要会议，即：两年一届在不同国家召开的世界生态安全大会和首届世界生态安全博览会的筹备工作；拟定于2011年6月会同联合国大学、中国国际问题研究基金会在中国举办气候变化与生态安全论坛。

二是与联合国机构、欧盟委员会合作，依据生态安全管理体系继续指导生态安全城市、生态安全社区、生态安全园区、生态安全旅游区、生态安全企业建设，对生态安全产品（项目）实施评估和认证。

三是继续对亚洲和非洲欠发达国家和最不发达国家进行技术扶贫和培训，以应对气候变化给上述国家带来的灾害和影响。

四是积极支持联合国大学《气候变化与生态安全》领域的课题研究，并参与联合国气候变化与生态安全大学的筹建工作，为联合国及亚洲政党国际会议、各国政党、议会、政府提供气候变化、生态安全与可持续发展方面的战略咨询。

五是抓好《国际生态与安全》杂志的出版，管理好世界生态安全大会与组织网站，进一步加强生态安全理论研究，开展学术交流与国际多边合作，发布年度生态安全报告。

六是抓好组织建设、制度建设，强化监督机制，使组织各项工作有章有序，实现科学管理，并寻求各国政府和国际组织的支持，创建"国际生态安全合作组织总部"。

三、几点建议与对策

第一，世界政治格局进入多极化之后，没有哪一种力量能够单独主宰世界，特别是一些关系全球发展的重大问题，诸如气候变化、森林植被遭到破坏、生物失去多样性、土地沙漠化、水资源污染、地质灾害频发等，这些问题仅靠某个国家和地区已无法单独控制，如何阻止地球资源的迅速枯竭，如何找到当代和下代人在经济社会和生态需求间的稳定平衡，怎样巩固发展中国家和发达国家的合作关系，是世界各国面临的迫切任务。我们决不能再以牺牲生态环境为代价换取一时的经济繁荣。面对当前频发的自然灾害与生态灾难，我们不能坐以待毙，必须建立跨国、跨境联动预警机制。一要加深对生态安全与环境保护理念的认识，各国应将生态安全与环境保护纳入全民教育体系。认真学习新西兰、日本的防震减灾经验和智利的防灾救援机制；二要建立国家生态安全领导与协调机构，从整体上解决部门间在气候变化、生态安全和灾害救援方面相互推诿、扯皮和官僚腐败问题；三要加强生态安全与环境立法，塞舌尔共和国与柬埔寨已先行一步把生态安全与环境保护确定为立国之本，并实施生态与环境立法。实践证明只有将生态安全与环境保护纳入法制化，才能有效遏制生态危机，真正实现生态、经济、社会的平衡发展。

第二，必须求同存异，加强国际多边合作，共同应对气候变化以及人为造成的灾害和冲突。从整体来看，世界不同国度、不同文化背景、不同发展程度等差异非常明显，促使各国合作进程必须面对一些难点问题，但只要各方本着"求同存异，和谐共赢"的原则，就有利于推进和平发展新格局的构建。当前生态危机总的形势是局部好转、整体恶化。因此，各国应在千年发展目标的大框架下，落实"共同但有区别的责任"这一基本原则。发达国家应积极主动承担维护生态安全和保护环境的责任，在实现和平发展方面发挥更大的作用；发展中国家应将生态建设与经济发展相结合，提高可持续发展能力，积极应对气候变化，并参与国际合作与竞争。

第三，当前，世界人口大规模流动和城市化快速发展的状况下，生态灾难随时可能发生。因此，要构建灾害应急机制，建立常设救援机构，通过各种有效途径提高政府、社会、民众对生态灾难的意识与紧急救援救助能力。我们将支持亚洲政党国际会议与马来西亚、日本等国家创建国际救援（亚洲）论坛和亚洲国际应急救援中心，并成立世界生态安全（森林恢复）基金。联合国千年发展目标将以一种前所未有的积极方式关注全球的发展与合作，我们准备为这些目标的实现作出新的、重要的贡献！

维护区域生态安全——理念与实践

中国工程院院士　李文华

一、生态服务功能是生态安全的重要保证

生态系统是生物圈中最基本的组织单元，也是其中最为活跃的部分。生态系统不仅为人类提供各种商品，包括食物、医药及其他工农业生产的原料等，更重要的是支撑与维持了地球的生命系统，包括涵养水源、保育土壤、固碳释氧、积累营养物质、净化环境、保护生物多样性、提供游憩和美学价值等。人们越来越认识到，生态系统服务功能是人类生存与发展的基础。据Costanza等（1997）的研究，全球生态系统每年提供的生态系统服务价值约为33万亿美元，为全球GNP的1.8倍。虽然对其计算的精确性和方法尚有争议，但这项研究在唤醒人们对生态系统服务功能的重视方面发挥了重要作用，并成为2000年联合国发起的千年生态系统评估计划（MA）的核心。

二、中国生态系统面临的挑战与机遇

中国自然环境先天不足，干旱、半干旱地区占国土面积52%，山地占国土面积2/3，青藏高原占200万平方公里，岩溶地区90万平方公里，黄土高原64万平方公里，脆弱的自然条件难以承受庞大的人口规模和快速的经济增长所带来的压力。

改革开放以来，经济的快速增长导致发达国家不同阶段出现的生态问题短期内在中国集中体现和爆发出来。近年来，中国的生态状况虽有所改善，但大多数生态系统仍处于退化阶段，对人体健康、经济发展、社会稳定乃至国家安全造成深远的影响，成为中国经济社会可持续发展的瓶颈。

据有关报道，目前中国水土流失面积356万平方公里，约占国土面积的37%；沙化土地面积约100万平方公里，近年来每年以3436平方公里的速度扩展；森林面积虽有所增加，但森林面积1.95亿公顷，森林覆盖率20.36%，仍低于世界平均水平（29.6%），且森林质量下降，林龄结构不合理，可采资源持续减少；90%的草地出现不同程度的退化，其中中度退化以上草地面积已占半

数，全国"三化"草地面积已达1.35亿公顷，并且每年还以200万公顷的速度增加；地下水超采严重，华北平原已形成全世界面积最大的地下复合漏斗区，达四五万平方公里，西部的许多地区，因地下水超采严重，大片已成活多年的树木枯死；生物多样性锐减——在《国际濒危野生动植物种贸易公约》列出的1121种世界性濒危物种中，中国有190种，外来物种入侵威胁生态安全，造成巨大的经济损失；环境污染已从陆地蔓延到近海水域，从地表水延伸到地下水，从一般污染物扩展到有毒有害污染物，已经形成点源与面源污染共存、生活污染和工业排放叠加、各种新旧污染与二次污染相互复合的态势。

三、中国在生态方面所做的努力及取得的成绩

另一方面必须指出的是，新中国成立以来，中国政府和人民在致力于生态建设方面做了很大的努力，投入了大量的人力、物力和财力，在保证国土的生态安全方面取得了举世瞩目的成绩。特别值得提出的有以下几个方面：

天然林资源保护工程。工程建设范围包括18个省区。国家通过天然林禁伐、限伐等措施，主要解决这些区域天然林的休养生息和恢复发展问题。调减木材产量1991万立方米；造林0.13亿公顷；管护森林1亿公顷；分流安置富余劳动力74万人。

退耕还林工程。工程建设范围包括25个省（区、市）和新疆生产建设兵团，共1897个县（含市、区、旗）。通过国家无偿向农民提供粮食和造林苗木等措施，计划到2010年控制水土流失面积0.23亿公顷，防风固沙控制面积0.27亿公顷。

京津风沙源治理工程。工程建设范围包括5个省（区、市）的75个县。通过大力封沙育林育草、植树造林种草、小流域综合治理、水资源合理开发利用等措施，计划到2010年林草覆盖率由6.7%提高到21.4%，主要解决首都周围地区风沙危害问题。

三北及长江流域等防护林体系建设工程。工程建设范围包括28个省（区、市）的1696个县。通过保护和增加林草植被，计划造林0.23亿公顷，管护森林0.72亿公顷。

生物多样性保护。中国目前已初步建立了生物多样性保护政策、法规体系和行政管理制度。同时，先后组织了多次全国性或区域性的大规模调查工作，加强了物种资源的调查与编目，组织和开发了生物多样性相关数据库。通过"就地保护为主，迁地保护为辅"的原则和办法，截至2008年底，中国已建立2538个自然保护区（不含港澳台地区），总面积14894.3万公顷，占陆地国土面积的15.13%，超过世界12%的平均水平，初步形成了类型比较齐全、布局比较合

理、功能比较健全的全国自然保护区网络。

湿地保护与建设工程。国家湿地立法工作取得了积极进展，形成了《中华人民共和国湿地保护条例》（草案）。截至2008年底，全国已建立湿地自然保护区550多个，湿地示范区面积221万公顷，全国拥有国际重要湿地36个，面积381万公顷。全国湿地公园总数达到80个，总面积59.6万公顷。

重点地区速生丰产用材林基地建设工程。以自然条件较好的南方地区为主，通过市场化和产业化的方式每年提供木材约1.3亿立方米。目前，重点地区速生丰产用材林基地已经形成了多种经济成分共同参与、多种经营机制并存、多元化发展的工程建设新格局。

石漠化综合治理工程。2008年先期在100个县启动了石漠化综合治理试点。工程建设范围涉及中南、西南8省（区、市）的451个县（旗、市、区）。总土地面积为105.45万平方公里，岩溶面积44.99万平方公里，其中石漠化面积12.96万平方公里。工程建设总任务为林草植被建设942万公顷，建设和改造坡耕地77万公顷以及畜牧基础建设。中央预算内投资4亿元，地方配套资金0.81亿元。

水土流失与生态安全综合科学考察。水利部于2005年正式启动水土流失科学考察，共有680人次参加，将全国划分成东北黑土区、北方土石山区、西北黄土区、西南石漠化区、北方农牧交错区、长江上游及西南诸河区、南方红壤区等7个区域，涉及25个省的292个县，累计行程13万公里，召开各个层次的座谈会260多次，采样1600多个，调查农户3200多户，搜集资料2100多份，照片5万多张，录像85小时，为摸清我国水土流失的现状和制定水土保持战略打下坚实的基础。

区域生态建设。20世纪90年代以后，中国掀起了以县、市、省为单元的生态建设的新高潮。中国的区域生态建设在科学发展观指导下，以不同尺度的区域单元为平台和切入点，积极探索在不同区域水平上保护生态环境、发展生态经济、创建生态文明以及建设新型社会，为我国社会经济的可持续发展做出了积极探索，在国际上也产生了良好的反响。

四、展望

上述工程的实施尽管取得了一定成就，但由于中国生态环境的脆弱性和历史"负债"的沉重性，中国在环境治理和生态系统保育方面的任务仍然是任重而道远。特别是近年来中国政府提出了以人为本、全面协调可持续的科学发展观，并对国际社会在造林绿化和节能减排方面提出了庄严的承诺。为了完成这些任务，加强生态系统的保育具有基础性和关键性的作用。以下几方面的工作，应给予优先考虑：

1．根据不同地区的自然——社会——经济状况进行因地制宜，分区施政，确定优先开发、重点开发、限制开发和禁止开发的原则，并按照不同生态系统类型进行分类管理。

2．对行之有效的生态保育计划应保持其连续性，以加强生态系统的保育和退化生态系统的恢复；继续进行已开展的生态保育项目。

3．在技术人才和资金方面有所加强。加强宣传教育，弘扬生态文化，动员全社会的参与，发展循环经济，减少生物资源的消耗。

4．完善相关政策和法令，逐步实现加强政策的探索，特别是在生态补偿并推进林权改革方面的体制建设。

5．加强有关生态系统服务功能方面的科学研究。

6．加强国际合作与交流。生态系统的影响超越国界，生态系统的保育也需要各国携手共进，特别是相邻国家的交流与合作尤为重要。借此机会，我们也呼吁东亚各国在生态保护和建设方面加强合作，促使区域的生态状况越来越好。

人类活动对热带疾病和痢疾的影响

卢森堡国际认证中心主席　皮埃尔·卢根

　　气候和文化行为一直都对媒介传播疾病和水传播疾病有重要影响。带菌体的成长需要特定的温度范围。在热带地区，气候变化和降雨为无数传染病的肆意繁衍提供了有利的环境。像蚊子一样，疾病载体在潮湿的季节数量更多。

　　作为环境和健康认证领域（ISO14000环境管理系列标准和OHSAS18000职业安全卫生管理系统标准）的专家，在国际认证方面，我们面临着热带疾病和痢疾的沉重负担。如果说恶劣的环境活动会对健康和疾病产生影响，反之更亦然：疾病和贫穷对环境、水污染、森林采伐、能源不当使用、滑坡都有灾难性的影响。

　　疾病可分为相对不同的两组。一组投入的防治资金充足，另一组资金投入不足。资金投入不足的如：利什曼病、疟疾、霍乱、登革热、恰加斯氏病、丝虫病、麦地那龙线虫病、阿米巴病、片吸虫病、颗粒性结膜炎、狂犬病、布路里溃疡、贾第虫病……相对于用于常见病的医疗费用，当前划拨到被忽视的疾病上的资金非常少。这些被忽视的疾病是数百万人死亡、残疾以及社会经济瓦解的主因。由传染病造成的生产力和教育机会的丧失以及高昂的医疗费用严重影响了无数家庭和社区。

　　每年有950万以上的人死于传染病——他们几乎都来自发展中国家。儿童特别容易受到传染病的侵害。肺炎、腹泻、疟疾是五岁以下儿童夭折的主要原因；脑型疟疾能造成永久的脑力损伤。

　　与许多国家和大学协会一起，我们集中全力与疟疾和痢疾作斗争。这些协会来自于非洲和南非洲。我们主要的赞助者是安塞乐米塔尔基金会和旋转卢森堡山谷。

　　与许多看法相反，疟疾一直都是温带国家甚至西伯利亚的常见疾病。"二战"后人们曾做出重大努力去消除这种疾病，并在大部分的北方国家如西西里岛、西班牙、佛罗里达、古巴、俄罗斯、安大略、阿尔及利亚、希腊等取得巨大成功。

双对氯苯基三氯乙烷（DDT）是治疗这种疾病的理想用品，但它的滥用引起了环境协会的关注，而且在贫穷的南方国家完全消除这种疾病之前，它就被禁用了。这是对充满戏剧化的后果的过度反应，因为尽管许多医学研究都试图证明DDT对人体的毒性，但迄今为止还没有任何研究能证明它对人体健康的负面影响。2006年9月，世界卫生组织解除了对DDT的禁令并把它用于室内滞留喷洒（IRS）。实际上，DDT对蚊子的排斥作用远远大于它的毒性。室内滞留喷洒为数百万可能死于疟疾的儿童带来了一线希望。

将DDT用于室内滞留喷洒能立即挽救许多生命，尽管我们欢迎这种使用，我们仍认为未来取决于草药产品而非化学杀虫剂。例如，印楝的提取物具有强烈的排斥作用和杀虫特性。

但是除了建立在蚊帐和杀虫剂基础上的预防措施，以草药为基础的疗法效果显著，因此会越来越重要。70%的世界人口仍旧依赖这种疗法。在这种新颖的方法中，一种植物发挥了重要作用：青蒿。在过去的几年里，我们积累的科学证据证明：如果在七天内连续饮用有这种植物提取的茶水（每杯水中2.5克），除了有复发的轻微风险外，疟疾感染就会完全治愈。饮用七天以上能减轻配子细胞血症和人与蚊子之间的传染。中国在两千多年前就认识到了这种植物的功效，而且没有发现人体对使用这种植物的疗法有任何的抗拒。然而在去年，青蒿综合治疗片却显示出了令人担忧的抵抗迹象。可能是因为它们缺乏干药草里的多酚的协同效应。

为了更好地协调我们的努力，我们在今年发起了"提倡草药医学的协会"（BELHERB）。它以来自于保加利亚和卢森堡的大学教授和医师们的工作为基础。几个世纪以来，中国和印度的草药医学创造了无数个奇迹。在分光镜工具的帮助下，我们想更好地理解它们的药物动力原理。

在我们卢森堡的研究工作中，我们发现青蒿茶对受污染的水有很强的消毒作用。事实是，把一杯青蒿茶倒入一升河水中能产生极好的饮用水。这种功效不仅被几家欧洲大学证实，还被塞内加尔、中非和哥伦比亚的大学证实。

哥伦比亚的安蒂奥基亚大学证实，青蒿对利什曼病和肝片吸虫有良好的治疗效果。

贝尔格莱德大学证实青蒿茶对选定的恶性细胞系如人类子宫颈腺癌、人类恶性黑素瘤Fem-x和BG、人类骨髓型白血病K562、人类乳腺癌MDA-MB-361和人类结肠癌LS174有细胞毒性作用。

在疟疾这一领域，我们的非洲伙伴达喀尔大学、班甘特大学、班基大学、雅温德大学取得了鼓舞人心的成果。你们可以询问他们的研究成果。

同时，我们在非洲和南美洲的许多国家都有种植园，并且种植园的植物药剂产品已经达到商业化水平。

　　BELHERB的科学家们充分认识到，在和南方国家学术机构合作的同时，还需要额外的研究工作来证实和扩展上述发现。如果这些发现得以证实，这将是极具开创性的工作。每天有20000名儿童死于疟疾、霍乱、痢疾、利什曼病……

　　气候变化对南方国家人民的健康有重要影响。艾属植物和其他的"中国"草药对他们来说是免费的药物，能够结束他们难以言说的痛苦。

发展绿色经济需要全力推进绿色标准化

中国山西省太原市市长　张兵生

尊敬的大会主席，
各位嘉宾，
女士们、先生们：

很荣幸能够参加本届大会，与大家进行交流。我认为，维护世界生态安全，既要在法律、政策、技术、战略等层面不懈努力，更需要在深层观念和经济理论层面实现绿色变革。

一、维护生态安全，需要推进人类由"经济人"向"生态人"的转变

工业文明取代农业文明，是人类文明发展的一大历史性进步。工业文明时代，人类的"经济人"特性得到了充分发挥。1776年亚当•斯密在《国富论》中第一次提出了以追求自身利益最大化为动机的"经济人"假设。我们应当肯定，作为"经济人"，人类追求正当的、必要的经济和物质利益是合理的。但十分遗憾，在现实中这种品性却走向了极端，漠视生态利益、过度追求物质利益和物质享受，最终导致了对自然生态等人类自身生存基础的严重破坏。维护生态安全，必须加快人类自身发展，对"经济人"进行扬弃和超越，由"经济人"发展为"生态人"。"生态人"是具有生态理性的人，强调在自觉尊重生态规律的前提下追求生态、经济和社会综合效益，倡导生态优先、全面协调可持续的绿色发展观。20世纪70年代以来，随着人类社会的发展进步，"生态人"作为一种全新的特性正在加速形成，"经济人"正在加快向"生态人"转变。在现实社会中，"生态人"体现为各种环保志愿者、绿色环保运动和绿色国际贸易、绿色经济与可持续发展浪潮、生态文明的发展等重大趋势。全球的生态安全水平由此得到提升。加速推进人类由"经济人"向"生态人"的转变，培育生态安全的新型社会主体，实现人性层面的自我革命、转型和提升，对维护生态安全具有根本性的影响。

二、维护生态安全，需要生态要素资本化和资本要素生态化，大力发展绿色经济

传统工业化模式之所以导致严重的生态危机，一个重要原因就是只把生态资源作为取之不尽、用之不竭的开发利用对象，而没有把它作为经济系统内在的资本要素，只强调物质资本的价值和作用，无视生态要素的价值和作用，将生态要素排斥在资本之外，即把生态系统与生态要素"外部化"。

如何破解这一现实问题？关键要从根本上克服物质资本和生态资本的对立，克服经济系统与生态系统的对立，实现两者的内在统一。生态资本化就是指生态系统作为整个经济系统的基础，从根本上参与了价值的创造过程，直接带来了更高的经济效益，并由此获得了重要的资本属性。同时，作为一种基础性的资本，它必须保值增值，人类的一切经济社会活动都建立在生态资本保值增值的基础上，都必须遵循这个规律。资本生态化就是指物质资本与整个经济系统的运行都必须尊重、体现生态价值和规律，特别是尊重生态资本保值增值的根本要求。在现实中，生态资本化和资本生态化具体体现为对传统工业经济和工业文明的生态化改造，推进由循环经济、低碳技术、清洁生产、绿色消费等构成的绿色经济的发展，实现从传统工业经济和工业文明向绿色经济和生态文明的大转变。

近几十年来，人类社会加速推进这种大转变、大变革，许多地方已开辟了提升生态安全水平的全新路径，生态安全形势出现局部好转。要维护全球生态安全，各国政府应当发挥第一推动力的作用，把绿色经济作为国家战略全力推进。当前，我国政府正在全力推进科学发展，加快转变经济发展方式。我所在的山西省和太原市两级地方政府也在推进经济社会绿色转型，大力发展绿色经济，并取得了积极进展。

三、维护生态安全、发展绿色经济，需要推进绿色标准化

随着传统工业文明向生态文明的历史性转变，支撑经济社会发展的标准体系，也正由传统工业文明的标准体系向体现生态文明要求的绿色标准体系转型。所谓绿色标准化，就是按照"生态人"绿色理念和绿色经济的要求制定实施标准并推进经济社会绿色转型的过程。它是把绿色发展理念、战略、政策等真正落到实处的关键性环节，也是将绿色经济变成人类自身活生生实践的极为重要的环节。为了生态安全，我们必须加快制定并实施切实可行、系统配套、科学合理、覆盖经济社会各领域和生产生活各方面的绿色标准，对全社会进行绿色标准的无缝化管理。有了绿色标准，不管人们是否真正理解它，但只要按照它去实践、去执行，就会获得最佳的秩序和社会生态效益。

　　运用绿色标准化来促进生态安全，也是国际上的成功经验。为了应对日益严重的全球能源和环境的挑战，我国逐步制定和实施了多项体现绿色经济和生态文明环保原则的法律、法规和标准。山西省作为我国高耗能、高排放产业相对集中的区域，近年来一直在积极制定和实施绿色标准，加快传统产业转型升级，推进循环经济、清洁生产和低碳技术应用，促进环境保护和生态修复。我们太原市，现在已经制定实施了45个绿色地方标准，成为我国拥有绿色地方标准最多的城市，有力推动了绿色生产生活方式的发育成长。

　　维护生态安全，需要人类从灵魂深处爆发绿色革命。让我们共同努力，加快推进和实现这场革命！

　　谢谢大家！

气候变化与生态安全 *

国际生态生命安全科学院院长　鲁萨克•奥列格

一、气候

"气候"这个词最早是两千多年前古希腊天文学家喜帕恰斯首先使用。希腊语中"气候"一词的翻译为"倾斜度"。

学者在这个概念里指出，气候是指地球表面对太阳的倾斜度，当时已经认为这是地球经度不同天气不同的原因。

气候是对某个地区多年天气现象的统计，取决于此地区的地理位置。和区域气候不同的是全球气候，它代表特定年限地球的整体气候。

图1展示的是地球气候系统：

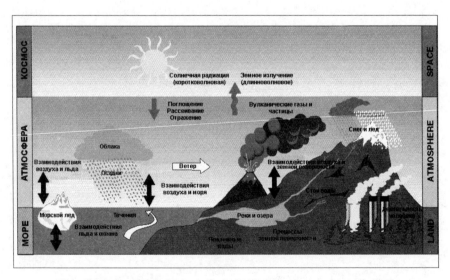

图1 气候系统组成部分，基本气候形成过程与相互作用图

气候形成是由太阳光线、大气压力、大气保温性、风向与风力、空气温度、海洋、基础表面与其他因素组成。

---

* 本文整理自演讲者的演示文档（PPT）。

而气候会变化是科学确认的事实。

图2表述的是地球22000年间的气候变化。

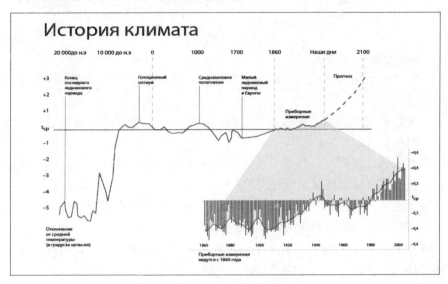

图2　地球22000年间的气候变化

图3显示的是北半球在最近5500年内年平均温度，可以看到，寒冷时期要比温暖时期长。

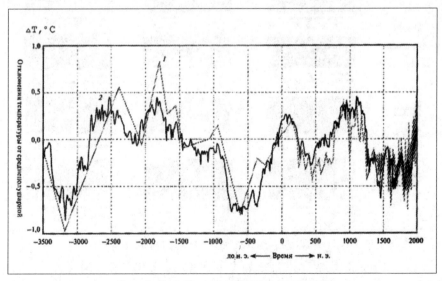

图3　北半球最近5500年间年平均温度

气候学是利用各种方法研究历史上气候的科学（树木气候学，叶绿素值，冰川气候学，历史文献等方法）。要理解现代气候变化，必须了解过去的气候现象。

二、气候变化的后果

伴随气候变化而来的是自然灾害现象。15000—20000年前人类度过了历史上最寒冷的时期。

正是在这个时期，人类取得了丰硕的成就——在所有大陆开始适宜居住，火的使用和语言得以发展等等。

那时虽然气候严酷，但那是一个卓有成果的时期，随之而来，进入了宁静的，甚至说难以察觉的黄金世纪时代。

气候的变化也带来文明与文化的变迁。这个地理位置决定理论，是由16世让·鲍丁创建，并由谷弥洛夫补充并发展。

欧亚大陆上，气候变化带来交替不断的干旱与洪水。历史上的气候恶化使民族迁移、王朝毁灭，也是收成不好、饥饿、疾病与瘟疫的原因。

图4为气候变化可能产生的后果场景图。

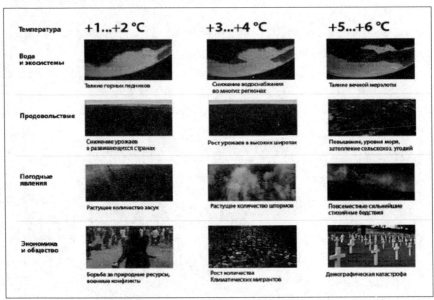

图4　全球变暖条件下可能带来的后果场景图
（斯戴伦发言稿，刊登于2006年《气候变化经济》简要）

现代气候变化信号在图5展示。

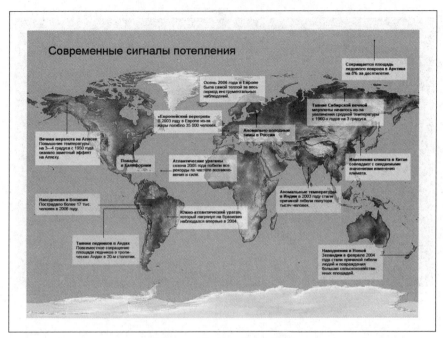

图5 现代气候变暖信号

（气候变化图，克里斯蒂·多，托马斯·多英，2008年）

三、气候变化的原因

近年来的数据显示，全球气候不断变化并带来对生态的不良后果。

气候学家继续讨论气候变化的原因，存在两种看法。

第一组：学者坚信，气候变化与宇宙以及地球上进行的所有过程有紧密的联系。如太阳的活性、火山活动、地球磁场极性变化等因素。

第二组：数量众多的学者都坚持，人类的活动是罪恶之源，就是说，人类因素的影响。坚信人为因素对气候变化的影响现在达到90%—95%。

气候变化的基本原因被称做"温室效应"，它是由空气中二氧化碳、水汽、氨气增多等因素引起的。

确实，从工业时代开始，甲烷在空气中含量增长到150%，二氧化碳增高了30%。温室气体在大气层中不多，约为1%。但它们在气候形成中起着重要的作用。

图6 地球"温室效应"产生过程

如果没有这些温室气体，地球气温将下降30度。

世界气象组织认为，工业革命前地球表面平均温度为13.7度，若没有温室效应则应为19度，但现在地表温度已达到14.5度。

我们认为，也存在第三种观点。当今气候变化是在自然与人类活动共同作用下形成的，其中人类活动的作用更明显。

四、全球变暖与其后果

从1995—2006年，根据自1850年以来的仪器观测结果，这段年份为最温暖的时期。

全球参数变化显示见图7。

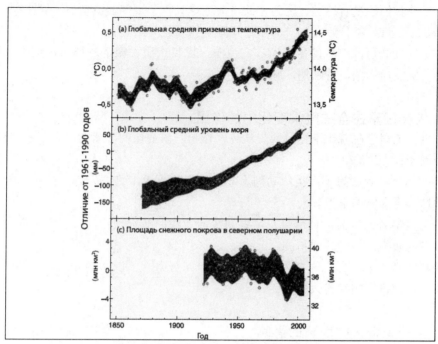

图7 气温、海平面、雪地面积变化图

气温、海平面、雪地面积变化可能产生的后果：

——干旱的概率增高；

——土壤的不稳定与滑坡；

——高山冰川的溶解；

——某些生态系统的消亡；

——某些动植物体系物种的消亡；

——水与食品质量的降低；

——气候难民的产生；

——疾病的产生。

五、即将出现的全球变冷

太阳是最重要的能量源。进入地球能量的数量取决于每200年一变的太阳射程变化。当太阳射程变小，则过15—20年，将会产生全球变冷，而当射程变大，则全球变暖再次来临。

太阳的面积在1980年末达到顶峰，1990年开始减少，意味着，将会有气温下降的现象。

目前我们还没有感觉，因为在此之前地球吸收了太多的热量。大约在2050-2060（+/- 11）年开始在45—60年内降温，然后又开始变暖。

全球变暖可能产生的疾病有：禽流感、巴贝虫病、霍乱、埃博拉疟疾、莱姆病、鼠疫、赤潮、昏睡病、肺结核、黄热病。

六、人类因素在全球气候变化中的作用

俄罗斯科学院院士卡特利科夫认为，影响气候变化的因素中，太阳占83%，人类因素只占2.5%。

大部分学者认为，人类活动因素对气候影响不大，许多人都支持宇宙物体对地球气候影响的学说。

一些学者们认为，目前没有足够的数据让我们准确预测，必须继续观测，必须客观研究，拒绝片面猜测。

科学必须朝解释原因并采取相应措施方向努力。可以影响并制约人类活动因素，但很难影响宇宙天体。

七、如何减低人类因素的影响

1. 寻找生态安全能源。
2. 清除烟气燃烧后的二氧化碳。
3. 对可再生能源利用率的增加（现在为4%）。
4. 潮汐与海浪能源的利用。
5. 风力的利用。
6. 太阳能的利用。
7. 地热及水热能源的利用。
8. 核能源的发展。
9. 可控的核热合成（俄罗斯已经进行了25年研究，但目前并不顺利）。
10. 生物能源的发展（木材废料），生物燃料的生产（生物粒料）。

八、减轻自然因素引起气候变暖威胁的独特理念

1. 人工火山。
2. 气溶胶硫在大气层中的雾化。
3. 环绕地球的反射镜（百万，d=60cm）。
4. 可反射太阳光线的盘状宇宙空间装置。
5. 培植吸收二氧化碳的海洋植物，并在海洋中喷洒有助于这些海洋植物生

长的物质。

    6. 极地人工造冰。

    7. 为控制树木砍伐，减少纸张使用（戈尔提议）。

    8. 其他的措施。

九、人类因素对气候影响中的国际关系

世界气象组织与联合国环境规划署于1988年共同成立了国际气候变化专家小组。由世界各国几千名学者组成。气候变化问题关系全球各民族与国家。

联合国对气候变化问题开展了规模宏大的工作。

1972年联合国大会在里约热内卢召开了环境与发展大会。世界156个国家的元首签订了气候变化框架公约，1994年3月21日开始生效。

现在已经加入公约的有包括俄罗斯在内的190个国家。

联合国气候变化框架公约针对减少温室气体排放共同努力，避免气候变化的危险。

必须保障排放指标稳定在1990年的水平。专家们认为，1990年的排放标准是相对安全的（见图8）。

图8 1990年二氧化碳气体全球地区排放结构

联合国气候变化框架公约是全球各界首次限制温室气体排放的尝试。但很快就表明，公约不能减少排放。1997年在公约中加入了《京都议定书》，为发达国家在法律上制定了在2012年之前减少排放的强制目标。

国际市场上出现了新的商品——温室气体排放配额。

　　俄罗斯1999年在《京都议定书》上签字，2004年批准生效。

　　美国2001年在《京都议定书》上签了字，但2001年拒绝对此议定书的批准。美国人口总数占全球6%，但却占25%的全球二氧化碳排放量。

　　到今天，各国对待《京都议定书》的态度并不明朗。

　　广泛认为，《京都议定书》只是政治性文件，而并非针对气候。

　　2007年巴厘举行的13次年度大会上通过截至2012年的新协约的制定。

　　2009年12月哥本哈根第15届气候大会不欢而散，毫无成果。

　　2007年英国在联合国安理会上提出气候变化特别议案。听证会举行了，但没有结果。2009年12月17日俄罗斯批准了气候报告，其中确定了针对全球与地区气候变化以及应对后果的国家政策。

　　美国总统巴拉克·奥巴马提出至2050年减少80%的排放数量。联合国秘书长呼吁把2009年作为"防止地球气候变化年"。

　　联合国认为，全球变暖是21世纪的首要问题。

　　最后指出，气候恶化的时代，我们必须加强人类思考，这将有助于我们在正确的道路上努力。

推进生态安全需要两大法宝

国际生态安全合作组织专家委员会主席、
中国建设部前副部长　杨慎

大会主席，

尊敬的各位嘉宾，

女士们、先生们：

根据大会安排，我就城市生态安全与灾害预警问题作一个发言，并借助刚刚闭幕的上海世博会这个舞台，阐发我对维护城市生态安全的基本观点。

为期半年的上海世博会于10月31日落下帷幕，这次以城市为主题的世博会共举办184天，参展国家和国际组织共246个，参观人数达7308万人，平均每天39万人。这是159年来第一次在发展中国家举办的博览会。

世博会闭幕的当天，由主办单位、世博局官员和参展单位共同协商形成的《上海宣言》向全世界广播，郑重宣布每年10月31日为"城市日"，呼吁人们站在历史的新高度，反思城市化过程中人、城市与地球家园的关系。宣言指出：城市化和工业化在带给人类现代化文明成果的同时，也伴随着前所未有的挑战。诸如：人口膨胀、交通拥挤、环境污染、资源紧缺、城市贫困、文化冲突等，字里行间折射出发展中国家面临的困境和对未来的企盼。

《上海宣言》最大的亮点是：它所阐明的维护生态文明、推进生态安全的理念，同本次大会的目标和主旨完全一致。《上海宣言》把"创造面向未来的生态文明"放在首位，充分表达了发展中国家及世界各国对加强生态安全的共同愿望。

女士们、先生们，作为一个致力生态安全的工作人员，我认为维护生态安全是全人类的神圣职责，加强灾害预警是大家的共同义务，建议当前重点从两个方面推进：

第一，加强理论体系建设。人类社会经过漫长的历史，到目前为止，大致经历了四个发展阶段，这就是史前文明、农业文明、工业文明和目前正在进行的科技文明。然而，科技发展是没有止境的。可以预期，不久的将来，人类很可能进入生态文明的时代。生态文明最基本的特征有二：一是社会生产力高度发达；二是人、社会与大自然和谐共处、和谐发展。这是全人类共同为之奋斗的目标，也是本届生态安全大会为之奋斗的目标。

因此，必须把普及生态安全知识，奠定生态安全理念放在最重要位置。在这方面，蒋明君先生编著的《生态安全学导论》为我们提供了一个入门的蓝本。我建议在全社会首先是各级公务员中，在中小学及各级学校中，在公共场合中，广泛普及生态安全知识，做到家喻户晓，人人皆知。同时积极创造条件，创办《国际生态导报》，开通国际生态网站，并通过电视杂志及各种媒体宣传普及生态知识，提升广大民众的生态安全知识水平。

第二，坚持依法推进。生产安全是利国利民的公益事业，必须动员和组织广大群众积极参与。但仅进行知识的普及还远远不够，必须在普及生态知识的同时，形成一整套的法律法规体系，依法治业。为此，必须加强立法，严格执法，并建议海牙国际法庭设立国际生态安全法庭，依法规范国家间的法律行为，在全社会、全球形成遵守制度的法律环境。

总之，教育和法制，是推进生态安全的两大法宝。以上是我的发言，祝大会取得圆满成功，谢谢大家！

三、专题论坛

（生态安全与灾害预警）

在资源稀缺中寻找最佳的发展模式

欧洲PA国际基金会秘书长 里约·普拉宁

我们的世界有约68亿人口，其中，大约9.25亿人每天都得不到足够的食物。导致这种现象的原因有很多，可能源自自然或是人为的危机。这些危机可能是系统性的，也可能是随机发生的。但它们都有一个共同特点，那就是存在着稀缺。

稀缺是一种经济现象。专家指出经济是稀缺的科学。的确，如果我们讨论今天的经济模式，或者是过去应对稀缺的那些成功或失败的经验，都是对经济政策措施的讨论。

在今天，全球化的经济政策以及由此而带来的稀缺越发明显且直接地受到来自本国和国际发展的影响。可能在过去，这些影响不如今天那样明显，但全球经济显然处于一个巨大的再调整过程中。这意味着筹资和投资比过去更趋向于流向投资收益好的地方。

外国直接投资、投资回报以及对稀缺的管理创造了新的压力，并引发了安全上的忧虑，在区域内和全球范围内皆是如此。了解这样的巧合是如何发生和演化，以及它对将来有何种指导意义是非常有用的。

除了政治/战略/意识形态的视角，预先准备和预防措施也能够缓和这些压力，但如果这些措施以增加防御支出/军费开支的形式出现，压力反而可能增加。

同时，北方国家对南方国家的官方发展援助正在减少，并且未来减少的趋势将会更加显著。受援国越来越多地被认为是贸易合作伙伴，那么为什么你要用一笔更好的交易来帮助你的对手呢？然而，在真正处于危急关头的人道主义问题的背景之下，稀缺仍然是决定性的因素。那么，如何将上述两种因素结合起来呢？

另一个同样令人不安的问题是官方的外交和防御政策与产业政策的关系，以及政府机构和产业的关系。同时，主权资金在全球的国民经济中扮演着日益重要的角色。最后，大量的全球贸易和投资表明企业社会责任应该有更强的基础，并且经济和市场的发展是在公民社会和产业相互利益得到充分的相互理解的前提下进行的。许多人认为企业社会责任应该成为公私伙伴关系的一种新形式。如果这样的说法是正确的，那么"应该在官方发展援助和企业社会责任之间建立起新的联系"这一说法也应该同样正确。

近来，我们又一次面临了这样的事实：汇率和量化宽松政策影响了贸易、就业、收入乃至税收、医疗、教育、国防开支、环境保护等各个方面。货币政策和保护主义也增加了人们的不安全感，以至稀缺感。

那么，这些财政、经济、发展援助、国防开支等现象是如何结合起来的呢？世界银行、国际战略研究所、世界卫生组织、联合国开发计划署等组织的统计资料，分析不同的资金流和数量随着时间推移而结合。当然这里面存在着不实之处，该死的谎言和数据，因此流动模型只是一种象征性的，试图使比较成为可能的东西。

上述问题说明，在世界的不同地区都存在着剧烈的变化。事实上可能存在着三种相互作用和变化的平衡：全球军事力量平衡，全球贸易平衡以及全球生态平衡。这三种全球性的平衡相互作用，并通过个体公民对安全的感知为在地球每个角落的公民创造了一种环境。当这些全球性的平衡影响个体公民的时候，这个人所能感知到的个人安全状况基于他应对稀缺的能力：这个人能否为他或她的家人提供足够的住所、食物，保障他们的健康和教育。

到目前为止，大多数公民是受聘于中小型企业的，他们的收入为家庭提供生计。今后，通过贸易、投资、社会责任、公私伙伴关系等，把中小型企业的发展和寻求提高中小企业的稳定性和成功能力的政策相结合，其重要性不言而喻。

如果上述所说的内容在相对和平、稳定以及没有人为灾害的前提下都是正确的，那么在有巨大灾害打击的情况下就更为正确。需要再次强调的是，自然灾害是无法避免的，但是公民社会能够为此做好准备。同样的道理也适用于后冲突时代的公民社会管理。事实上，准备的程度以及预防此类灾难的措施对于个体公民处理不幸后果的能力而言是至关重要的。

约两年前，我们参与预防实施并且防止了一个禁止所有从亚洲国家出口的鱼类进入欧盟的禁令，这一禁令如果实施将是灾难性的。公私伙伴关系、企业社会责任方法和官方发展援助的结合不仅防止了一纸鱼类出口禁令，还创造了使这种鱼类能够向欧盟大量扩张出口额的环境——因而增加了就业，提高了个人收入，等等。

最后，我们可得出如下结论：

① 全球在财政、经济、军事和环境方面的再调整过程中存在剧烈的稀缺，需要深度地分析以找到潜在的更好的国际间的解决办法，同时避免保护主义。

② 全面观察政府和产业之间在全球化经济时代的互动要求，学习如何通过公私伙伴关系融合官方发展援助和企业社会责任路径的方法。

③ 全球变动中的军事、经济和环境的平衡以及由此产生的个人安全的概念这二者之间的互动，需要通过提高对中小企业的稳定和重视来重新审视。

④自然灾害无法避免，但公民社会能够为此做最优准备。个人安全的概念可以通过灾难和危机事件发生之后有准备地建设和恢复中小企业的措施中展示出来。准备和预防可能成为国家和国际发展援助、紧急救助和企业社会责任整体路径的一个重要组成成分。

找到人与自然和谐发展的最佳方法

联合国大学教务长　威瑟林·波波夫斯基

　　首先，我要感谢的是蒋明君教授为今天以及昨天的大会所做的一切。昨天的大会是首届世界生态安全大会，迈出了历史性的一步。这次大会把我们带到了当今许多重要的热点问题的核心。

　　我想要告诉大家的是为什么自然灾害不再是自然灾害，或者不仅仅是自然灾害，我把它们称之为非自然灾害，因为人类对自然的影响已经达到了破坏的程度。我们生活在一个需要界定人与自然关系的时刻。从历史角度讲，人类发展过程就是保护人类不受大自然灾害影响的过程，比如，洪灾、地震等灾害。而现在是保护自然不受人类破坏的时候了，人类破坏地球但也有能力保护地球。

　　前面鲁萨克教授就气候变化发展史和自然如何平衡人类生活做了精彩的陈述。现在的问题是人类是否也可以平衡生活。我们需要问一下这个问题。如果数百万年来都是自然保护人类的话，那么人类能保护地球吗？什么是生活的平衡因素？现在我们要谈谈保护生态安全以及环境的责任问题。我很欣赏刚才蒋明君总干事陈述的四点结论，因为我们通过研究发现，需要采取整体分析的方法来共同行动。

　　我们需要公共部门——由政府、民间团体组成；私人部门——由企业法人组成；但是个人、企业家和消费者也起了主要作用。因此，我们基本有四类行动者和四种保护生态安全的责任，有政府、民间团体、商业机构、个人、公共机构、政治家、私人机构以及企业家。那么，为了保持人与自然平衡，科学应该做些什么呢？最担心犯难的是决策者，他们总是要保持话语一致，要赢得选举。他们最先考虑的是经济发展，权力的最大化，财富的最大化，追求更大程度上的生产与消费。最大化的语言恰恰与生态安全的要求相悖。

　　我们需要谈一谈最大化的相反面，消费量与生产量更低，低碳节能，甚至计划生育。我希望政治家们的话语能从越来越多转向越来越少。一直以来人类总是将幸福与豪华轿车、大家族、豪宅、高档饭店以及设施便利的大城市联系在一起。但是这些恰恰是与未来需要相抵触的。

　　未来需要小型家庭、小型轿车，低需求，低消费，街道上的设施更少。我想，科学领域已经找到了某种方法。中国、俄罗斯、东南亚以及南亚国家在学术研究方面都做得很好。科学家们已经就某些问题给出了答案。我们需要要求人们改变生活方式。我们可以给人们提供绿色技术、低碳节能的生活方式，这样的生活同样能让人们感到幸福。

　　在此，我认为最重要的是亚洲政党国际会议与世界生态安全大会在同一个地方举行，可以真正地向决策者们传达信息。亚洲的生活方式一直是很和谐的。美国与欧洲认为的幸福的方式是建立在资本主义，建立在过度消费与消费主义之上的，正是这样的理念造成了我们现在面临的问题。而亚洲的幸福模式正好相反，它不注重资本主义，不追求权力财富的最大化增长、过度消费，而是更注重如何幸福和谐地与自然相处。我也目睹了亚洲科学界首先影响了其决策者们的决策，其次他们也向全球表明了他们找到了与自然和谐相处的方法。其中的一个方法是，未来的决策者们为赢得选举，会告诉选民们不要追求多生产、高消费、生育很多孩子、购买高档轿车、置豪宅，而是选择最小化的生活方式。

　　谢谢大家！

以"生态智慧"应对城市灾害与冲突

中国国际问题研究基金会副理事长　董津义

尊敬的各位嘉宾，
女士们、先生们：

大家好！非常高兴受邀出席本次世界生态安全大会，并作主题发言。围绕"气候变化与生态安全"这一主题，我想着重谈谈如何以"生态智慧"应对城市灾害与冲突。

生态安全是可持续的城市化的重要组成部分，也是近年来国际合作与交流中的一个热门课题。面对城市化进程中形形色色的灾害与冲突，我们需要以新的生态安全理念，指导城市的可持续发展。

地球上的自然变异（包括人类活动诱发的自然变异），无时无刻不在发生，当这种变异给人类社会带来危害时，即构成灾害与冲突。灾害与冲突是人与自然矛盾的一种表现形式，具有自然和社会两重属性，是人类过去、现在、将来所面对的最严峻的挑战之一。世界范围内重大的灾害与冲突包括：旱灾、洪涝、台风、风暴潮、冻害、雹灾、海啸、地震、火山、滑坡、泥石流、森林火灾、农林病虫害等。

在过去，一国发生重大自然灾害事件一般被认为是当地性的，影响也是有限的，防灾减灾是一国自身的事情。在经济全球化、全球信息化使各国相互依赖、联系互动愈益紧密的背景下，一国或一个地区发生重大自然灾害将不可避免地"殃及池鱼"，给国家稳定、发展与安全带来负面影响。比如，重大自然灾害会导致巨额经济损失和政府财政支出，造成民众严重的心理恐惧和社会恐慌。如果灾害应急和后续救援不得力，还有可能触发民众骚乱、社会动荡，甚至造成政治危机而影响国家稳定。另外，重大灾害的发生迫使一国把注意力集中于国内，限制了同时应对涉及本国安全的外部重大事件和危机的能力。

仅今年7月中旬以来，受炎热酷暑的袭击，俄罗斯接连引发大规模的森林火灾，森林着火点多达550多个，泥岩着火点20多个，着火总面积超过19万公顷。持续不散的烟雾致使莫斯科空气污染物含量达到有史以来最高，日死亡人数在

700人左右，同期相比上升1倍，民众一度感到恐慌。7月下旬以来，由于连续降雨，巴基斯坦遭遇了80多年来最严重的洪涝灾害。据估计，洪水摧毁了大量的民宅、公路、桥梁和农田，造成了至少1600人死亡，1500万人受灾，400多万人无家可归。2010年8月8日凌晨，中国甘南藏族自治州舟曲县突降强降雨，县城北面的罗家峪、三眼峪泥石流下泄冲向县城，造成沿河房屋被冲毁，泥石流灾害造成1200多人死亡，588人失踪，2万余人受灾，人员伤亡人数相当于过去三年全国因滑坡、地面塌陷、泥石流等种类地质灾害死亡和失踪人数的总和。

　　自然灾害是难以完全避免的，虽然人们不断地与自然灾害做斗争，采取各种防御办法，但灾害犹存。灾害不可完全避免并不意味着人们在灾害面前就无能为力、无所作为。恰恰相反，我们必须发挥人的主观能动性，以"生态智慧"应对城市灾害与冲突。

　　"生态智慧"的思想，最早产生于中国。所谓"阴阳两极"、"天人合德""道法自然"等思想，是古代中国人在与自然界相处过程中努力寻求秩序与和谐、寻求与自然共存共荣的高度智慧结晶。当前，在灾害频发、祸害不绝，人们密切关切、讨论防灾减灾的时候，有必要重新审视古老又先进的"生态智慧"，用来探索城市的生命历程及其栖息的外界环境系统之间的互动关系，合理布局，科学发展，使我们的生活、生产与自然的生态达到高度的一致。为了更好地应对灾害与冲突，尽可能避免灾害与冲突的发生和减少其带来的损失，我们应当做好以下几个方面。

　　1. 进一步提高防灾意识。各国政府应当加强科普宣传教育工作，提高全民的防灾、减灾意识，通过各种途径积极开展防灾、减灾的宣传，普及教育，提高公众的生态安全意识和减灾意识。

　　2. 完善应急管理机制。灾害与冲突的发生通常有一定的征兆，这种征兆可能会被人们通过科学的监测手段获得，采取措施，避免灾害的发生。对于已经发生的灾害与冲突，则应当尽早采取措施进行应对，控制灾害，防止灾害的进一步扩大、蔓延，以尽可能降低灾害造成的损失。因此，一旦灾害与冲突发生了，就应当对灾害与冲突进行积极迅速的应急处置。

　　3. 加强灾害与冲突应急预案。灾害与冲突应急预案是灾害与冲突应急管理的重要内容，是灾害与冲突应急管理中不可缺少的重要环节。灾害与冲突一旦发生，有关部门则可以根据预案并结合实际情况采取相应的措施，对自然灾害进行处置。

　　4. 做好灾害与冲突应急管理工作。灾害与冲突一旦发生，短时间内可能需要大量的人员、救灾物资和资金。因此，需要在平时做好应急管理，这包括专业人员、救灾物资和救灾资金等方面的储备工作。

　　总之，城市的灾害与冲突往往具有突发性、无序性、危害性。政府之外的不同层级部门机构也应当积极做好防灾、减灾工作，还需要动员社会各个方面共同配合，甚至加强国际合作，共同应对灾难，建立科学的防灾减灾工程，从源头上控制自然灾害的发生。

在治理荒漠化中的生态安全问题

俄罗斯国际商业技术大学校长   乌格德奇科夫

尊敬的女士们，先生们：

我非常感谢首届世界生态安全大会主办方给我提供这次发言的机会。在此我简短介绍一下治理土地荒漠化情况。2005年11月4日，由政府组办的俄罗斯联邦城市科学观察委员会建立开始，我就担任此委员会位于米丘林市的委员，并成为前苏联与俄罗斯可控生物合成技术科学知识领头人（微生物、动物、植物可控细胞培养），这个研究方向的奠基者之一为波罗金院士（1921—1999）。

俄联邦城市科学观察委员会的创建目的，主要是研究农业新技术并向全俄范围推广。

俄联邦城市科学观察委员会建立了可控生物合成科学实验基地：

——可控细胞培养实验室；

——批量工业生产线；

——农业作物培育及有机垃圾处理试验田。

在我所担任总干事的波罗金国际生物技术基金会，完成了专业生物技术设备（电子生物反应器）的研制。

因为拥有生物技术科学生产基地，我们成功研制了沙土地蔬菜与景观。

植物生物技术溶液，对联合国制定的2010—2020年度治理沙漠计划执行有非常重要的帮助。同样重要的是，这些生物技术不使用任何化学试剂（废料、植物防护物质等等），保障了土地生产的有机性。此技术2010年在俄罗斯获得专利。无需化学试剂的作物的培育保障了高度的生态安全性。

值得关注的是，在俄联邦城市科学观察委员会生物技术科学生产基地，还可用生态安全的工业生物技术处理残余食品垃圾，把其变为培育植物的高效生物肥料，包括在沙土（沙漠）的条件下。可同时处理50家大型食品加工企业的食品垃圾。

生物技术处理后的垃圾可以用在沙漠地带作为土壤厚层，经这种处理方法后的垃圾可代替一般使用的昂贵的、较少的泥炭与腐泥，作为沙漠地带条件下的土层。这个项目被称作"垃圾与效益"完全合理，因为食物垃圾的处理极大地改善了城镇的生态安全，也减少了在治理沙漠化过程中创建人工土壤层的物质消耗。

俄联邦城市科学观察委员会的另一个重要成果是对加快蔬菜、景观植物种籽萌芽速度的生物技术溶剂的研制开发，在荒漠及半荒漠、风蚀土壤严重的地理条件中，对植物的生长尤为重要。

可以看到，在俄罗斯以及在那些关注联合国荒漠治理计划实施的国家创建这样的科学生产基地（中心）极有意义。俄联邦城市科学观察委员会主席（米丘林市市长、农学博士马卡罗夫）与联合国工业发展组织莫斯科代表处主任（卡拉特科夫）签订的《工业发展组织系统框架下在农业与环境保护领域中推广生物工程技术计划》（2009年8月14日）的项目，就朝着这个方向迈出了第一步。

为了加快新技术的推广运用，国际商业技术大学于2009年11月设立了国际跨越创新技术学院，把多项科学交集（物理、化学、生物等）使用在不同行业上：新材料、机器制造、交通、生物技术在农业、医学及环保领域中的工业化。国际跨越创新技术学院院长马伊吉斯教授邀请有兴趣的各方共同合作。

作为结束语，我站在如此高层次的讲台上向大家提出建议。就像前面所说的，在联合国未来十年荒漠化治理项目实施的框架下，将会推进大量的国家或多国合作项目。

我提议在我们的下次会议中，讨论举行国际荒漠化大会的议题，邀请所有感兴趣的国家参与，预计在2012—2013年召开。

感谢在座的关注。

把凌云山建成国际生态旅游的最佳范例

中国四川省南充市纪委书记　胡文龙

尊敬的各位来宾，
女士们、先生们：

应邀来到美丽的柬埔寨金边市，参加世界生态安全大会首届年会，我们感到非常荣幸。随着全球人口增长和资源开发利用，人类活动对环境的破坏性不断增大，生态安全面临日益严重的威胁，已经引起世界各国政府和国际组织的广泛关注。近年来，虽然世界各地在预防自然灾害和应对生态灾难上措施得力、成效明显，但环境逆向演化的趋势并未从根本上扭转，由环境退化和生态安全危机引发的环境灾害和生态灾难并没得到有效遏制。因此，不管是个人，还是区域和国家，都面临着来自生态环境的严峻挑战。保持全球及区域性生态安全已成为国际社会的普遍共识。大家越来越深刻地认识到，只有加强合作，资源共享，维护生态安全，才能实现互利共赢。

南充市位于中国西南、嘉陵江中游，总人口756万，幅员面积1.25万平方公里，素有"千年绸都"的美誉，被誉为嘉陵江畔的一颗璀璨明珠。嘉陵江是长江水系最大支流，是南充的"母亲河"，流经南充7个县（市、区），长达298公里，沿江两岸绿意盎然、万鸟竞翔，生长着各类野生植物2000余种、野生动物665种。嘉陵江流域生态文化旅游区是四川省在"十一五"期间重点建设的"新五大旅游区"之一，嘉陵江南充段是其核心区域。凌云山景区依江而立，是嘉陵江流域开发建设的重点项目，目前已成功创建为中国AAAA级景区和国际生态安全旅游示范基地。在凌云山景区开发建设过程中，我们始终坚持人与自然和谐、共融共生的发展规律，走保护与开发并重、可持续发展之路，通过实施退耕还林、恢复生态、挖掘历史文化，将城市近郊的荒山规划和建设成为一个山水交相辉映，人文相互依存，具有生态、文化和宗教特色的生态安全旅游项目，高度体现了生态文明与生态安全理念，探索出经济、生态和社会"三位一体"的发展模式。

保护与开发并重，丰富凌云山生态资源。景区森林覆盖率达86%，植物种类繁多，乔灌木树种多达千余种，野生珍稀动物近百种。山水为形，神奇为魂。凌云山景区二佛（72米卧佛、99米立佛）并列，三山（白山、图山、凌云山）相连，四海（林海、云海、花海、竹海）缤纷，五湖（凌云湖、白山湖、图山湖、玄武湖、小西湖）环绕。这里不仅是国家级森林公园、省级地质公园，还是四川省文化产业示范基地、中国最佳风水旅游景区。丰富的原始森林资源是凌云山旅游区最重要的自然遗产，也是创建凌云山国际生态安全旅游示范基地的重要基础。

传承与弘扬齐奏，发展凌云山宗教文化。儒教、道教、佛教"三教"共存共生，在凌云山景区得到完美诠释和集中体现。自汉唐以来留存在凌云山大量的儒、佛、道三教寺院庙阁，至今仍保存完好。文献记载的"三月三，朝灵山"重大法事活动至今依然盛行。这里的宗教文化源远流长，具有重要的保护和研究价值。

心灵与自然感应，共享凌云山风水文化。以凌云山主峰为中心，周边群峰在山貌形态上形成了古代传说中的龙、虎、龟、雀巨型"四灵兽"之相，与中国古代风水学所称的"四相五行"玄机契合，而且自然按风水学方位各占其位（前朱雀、后玄武、左青龙、右白虎），形神兼备，惟妙惟肖，世间仅见，被称为中国风水地理的活标本，研究价值极其宝贵。

今年3月，国际生态安全合作组织授予凌云山景区"国际生态安全旅游示范基地"称号，是对景区近年来坚持实践经济、社会与生态协调发展理念的充分肯定。希望国际生态安全合作组织继续关注凌云山的开发与建设，在资金、技术、宣传等方面给予大力支持，力争早日把凌云山景区建设成为自然遗产保护与生态、经济、社会协调发展的最佳范例。

尊敬的女士们、先生们，南充是中国四川东北部区域中心城市，因生态建设吸引着世界的目光，也因生态建设与国际生态安全合作组织结缘。我们热忱欢迎各国政要和国际友人到南充考察访问，欢迎各国的实业家到南充投资兴业，加强经贸、生态等方面的合作，共同开创美好的未来。

最后，祝愿本次大会圆满成功！

谢谢大家！

全球生态安全与灾害预警

加拿大列治文市教育委员会副主席　区泽光

在全球生态安全问题日益严重的环境下，国际间的合作比以前任何一个时期都更重要，因为生态安全问题不是只影响个别国家和地区，而是广泛地、整体地威胁每一个国家。生态环境被破坏所造成的灾难，往往是扩散的，并会造成持久的影响。以美国2010年4月发生的墨西哥湾深海钻油台漏油事件为例，受威胁的并不局限于美国本土和附近海域，这场历时超过3个多月的特大漏油事故在被截停之前，已漏出1.85亿加仑石油，在海床沉积了厚达2英寸的油污和沉淀物，杀死了约30%的渔产并威胁400种生物，至今专家们仍在努力评估生态损害的程度，但可以肯定全球的海洋生态必定大受影响。

虽然不是每次生态灾难都达到同样的恶劣程度，但仍会产生跨国的危机。在2005年11月由中石油吉化公司双苯厂爆炸所造成松花江水体污染事故，就让俄罗斯大为紧张，恐怕毒水会流入境内，也凸显了国际上预警和通报合作的重要性。这个例子足以说明，一些较轻微的事故，也可以随着空气、水流和地下水跨越边界把影响扩大，当位于河流上游的国家发生生态安全事故，将无可避免地影响下游国家和地区，如果事故的严重性最初没有引起注意，而下游国家的监察系统不完善，或有害物质已稀释至一个难以侦测的浓度，下游国家的人民可能成为无辜受害者，甚至导致国与国之间的纷争和冲突。

尤其值得注意的是某些轻微事故，最初或没有实时影响，但如果掉以轻心，任由它产生累积效应和延迟效应的话，后果可能非常严重。例如少量的有害化学物质不断渗入泥土中，后来随着地下水向外扩散，受影响的地区初时或许很难察觉，一旦累积到可被发现的程度后，且已错过了防范的机会，将造成生命和经济的损失。

尽管生态危机迫在眉睫，令国际社会不能不正视，但政府之间的合作仍存在很多障碍，其中一例是各国迄今仍未能达成有约束力的协议去减少碳排放量；另一个例子是国际上对于生态安全事故仍未建立起一套标准化和有效率的通报机制，可以在事故发生后能及时发出警报。正是因为缺乏一套规范化的通报制度，所以当墨西哥湾漏油事故发生后，英国石油公司（British Petroleum，BP ）被指延误及隐瞒通报，令美国及国际社会未能在第一时间掌握正确的情报，从而采取适当的相应措施。要制止生态环境继续恶化，最彻底的当然是通过国际合作，消除引致环境恶化的根本因素。但透明而有效的预警通报机制也是不可少的，因为大小的生态安全事故，可说无日不发生，事故一旦发生，及早防范它的蔓延，把灾害减到最小，可说是刻不容缓的。

在国际上为了保障共同利益而建立通报制度，其实不乏先例，例如国际原子能机构（IAEA）在1986年制定了《及早通报核事故公约》《核事故或辐射紧急情况援助公约》和一系列与核安全、辐射安全、废物管理安全标准有关的国际公约，有效地加强了核能方面的安全发展和国际合作，并且在缔约国之间建立有关核事故的通报机制，使可能超越国界的辐射后果减至最低限度。另一方面，世界卫生组织（WHO）多年来促进及完善国际传染病疫情的通报、协调和信息交流，已证明有效预防和控制疾病的流行，对保障人类健康作出重大贡献。这都是加强国际生态安全合作值得借鉴的例子。

要建立国际生态危机的通报和预警机制，最先决的条件是国与国之间能建立信任和互助的基础，各国必须明白大家是祸福与共，既然生态事故的影响容易扩散，某个国家很难说可以置身事外。国与国之间的信任和合作最少包括三方面：第一，当事故发生后，有关国家必须以负责任的态度，及时向邻国以至全世界发布消息，减低风险扩散的机会，不应任意以国家机密为理由造成延误；第二，如果有需要的话，应接受外国的支持，共同研究对策，尽早解除危机，减少破坏和损失；第三，在事故结束后，整理有关资料与其他国家分享，希望通过交流经验，提高各国预防同类事故再发生的能力。

各国间的互信是需要通过接触、对话和了解然后产生的，因此，一些高层次的国际交流平台，如国际生态安全大会或类似的峰会将有助交流，从而建立共识及长期合作基础。不过，由于国家之间存在地理、文化、政治、经济发展阶段等差异，要所有国家能同时达成一致意见，必定需要时间和耐性方可成事，一个可以考虑采用的模式是争取在一个较小地区范围内，由条件比较相近的国家建立地区性合作，然后逐步融合扩大合作区域和范围。

在政治层面以外，要建立国际危机通报机制，有一些技术问题是需要解决的，其中最重要的是风险评估的标准化，由于生态安全事故的多样性和复杂性，各国必须在评估标准时采用一致的准则，例如按事故的性质、种类、危险程度、扩散风险等各方面，建立不同评估等级，然后确定通报的时限和层次。如此一来，事故一旦发生，就可以迅速评估形势，并根据风险的级别启动不同程度的通报机制，这样才可以发挥预警的效用。如果风险评估标准不统一的话，只会造成混乱，效果适得其反。

当然，要建立一套完备的国际化风险评估标准，是一件富有挑战性的工作，有赖各国的研究组织、专业团体和学术机构通力合作，才会成功。

最后，各国也要完善本身的监察系统，因为如果本身的监察系统存在漏洞，无法对境内的生态安全事故进行有效监测的话，国际的预警机制也是形同虚设，不能发挥应有作用，所以，发达国家应在这方面对资源不足的发展中国家提供援助，这是利己利人之举。

全球生态安全问题毫无疑问已达到了极其严峻的阶段，国际合作若仍停在空谈的层面，极可能会错失最后的关键性时机，唯有透过长期和全方位的合作，建立完善的监察及预警机制，才有望扭转形势，使我们的天空重现蔚蓝，大地再展生机，绿水永远长流。

谢谢！

关于塘厦生态建设的实践与思考

中国东莞市塘厦镇委书记、人大主席　叶锦河

尊敬的各位来宾、朋友们：

大家好！

冬日金边，景色秀美。首届世界生态安全大会在这里隆重召开，我谨代表中国东莞市塘厦镇委、镇政府对大会召开表示热烈的祝贺！借此机会，我也真诚地感谢主办方：柬埔寨王国政府、亚洲政党国际会议、国际生态安全合作组织为我们出席本次盛会给予的支持和帮助。

塘厦镇地处中国南部的珠江三角洲，介于东莞—深圳—香港经济大走廊之间，是广东省中心镇、东莞市东南部区域中心，全镇总占地面积128平方公里，总人口40多万。沐浴中国改革开放的春风，塘厦镇经济社会飞速发展，从一个普通的农业小镇，发展成为一个国际加工制造业重镇、世界高尔夫名镇、中国经济发展千强镇第五名，去年全镇GDP达168.7亿元，预计今年GDP将在去年的基础上增长10%。

纵观塘厦镇工业化和城市化的发展历程，我们始终坚持"经济与环境双轮驱动、人与自然和谐相处"的科学发展理念，在全力推动经济发展的同时，高度重视自然生态保护和社会公共事业发展，大力实施"绿色GDP工程、蓝天工程、碧水工程、宜居工程和绿地工程"，优化园林规划布局、提高园林绿地配置，发展低碳循环经济、提高城市环境质量，落实节能减排增效、提高城市综合承载力，加强社会综合治理、提高幸福生活指数，从而构筑了塘厦生态绿地环绕、绿色经济循环、和谐文明共存的动态安全生态体系，打造了"城在林中，人在绿中"的岭南特色园林城市，促进经济社会与自然环境的协调持续发展，营造宜居宜商的良好环境，实现了"既有金山银山，又有绿水青山"的城市发展规划目标。

目前，全镇建成区绿化面积达2318.02公顷，占镇域面积的20%以上，人均绿化面积达15.6平方米。横卧在城市中央的绿色道路长廊——迎宾大道的四季常青，矗立在城市西南面的大屏障森林公园构成城市天然氧吧，大钟岭湿地公

园点缀着塘厦原生态山水园林，被吉尼斯认定为世界最大的塘厦观澜湖高尔夫球会，更映衬出人居环境和自然环境的高度和谐统一。

塘厦镇城市生态安全建设取得可喜的成绩，先后荣获"中国国家园林城镇"、"中国绿色名镇"、"中国国家级生态乡镇"、"中国卫生镇"等称号。今年7月30日，经过联合国国际生态安全科学院和国际生态安全合作组织的专家评估，塘厦镇被授予"国际生态安全示范镇"荣誉称号，这标志着塘厦"生态安全"建设迈上新的台阶。

通过塘厦生态安全与可持续发展的实践与反思，我们认为，生态安全城镇建设是一项综合性工程，它是一个文明、健康、和谐、充满活力的复杂系统，是生态良性循环的区域形态，是一种人与自然和谐共处、同气共生的境界。

从生态哲学的角度看，生态城镇的实质是人与人、人与自然和谐相处。塘厦的生态城镇建设特别强调人是自然界的一部分，人必须与自然实现整体协调，在和谐的基础上实现自身发展。整体性是塘厦生态城镇建设的价值取向所在。

从生态经济学角度看，生态城镇既要保持经济的持续增长总量，更要保证经济增长的质量。生态城镇倡导绿色能源的推广和普及，致力于可再生能源高效利用和不可再生能源的循环节约使用，关注人力资源的开发与培养。持续性是塘厦生态城镇建设的强大动力所在。

从生态社会学角度看，生态城市不是单纯的自然生态化，而是人类生态化，即以教育、科技、文化、道德、法律、制度等各方面的全面生态化为特色，建立自觉保护生态环境，促进人类自身发展机制和公正、平等、安全、舒适的社会环境。全面性是塘厦生态城镇建设的根本追求所在。

从地域空间角度看，生态城市不是一个封闭的系统，而是以一定区域为依托的社会、经济、自然综合体，在地域上生态城镇是个城乡结合体的概念，即城镇与周边关系趋于整体化，形成城乡互惠共生的统一体，实现区域可持续发展。兼容性是塘厦生态城镇建设活力永葆所在。

朋友们，人与自然和谐共生既是我们共同的责任，也是我们共同的追求。我们将以荣获"国际生态安全示范镇"为新起点，努力把塘厦建设成为和谐幸福秀美的现代化新城，为我们地球的绿色家园更加秀丽作出更大的贡献！

祝大会取得圆满成功！

热情欢迎各位来宾、朋友们莅临中国东莞塘厦投资、做客！

高效益危机管理中的合作伙伴 *

全球发展媒体视点主任　奥托•伊伍让斯

广播媒体
高效益危机管理
中的
合作伙伴

广播媒体—社会发展的合作伙伴　　　　　　　　　1

---

* 本文整理自演讲者的演示文档（PPT）。

以往的经验告诉我们：

- 危机发生之前、之中和之后，对信息的规定、处理和格式设置对于为个人和社会部署干预的质量和结果是至关重要且相互依存的。
- 有效调度信息资源并专业开发使用这些资源的系统有待进一步改善。

广播媒体-社会发展的合作伙伴　　　2

鉴于经验，我们已经展开以下行动：

- 开发一个国际化、高质量、可以本地化的信息资源库。
- 利用广播技术强化对于受众来说价格低廉且充足的信息渠道（推广视频点播）。
- 改进专业媒体技能，以便在危机中提供由媒体支持的干预。

广播媒体-社会发展的合作伙伴　　　3

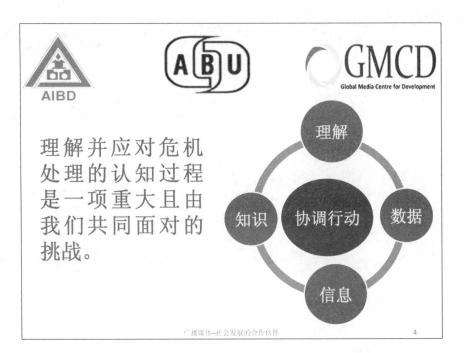

理解并应对危机处理的认知过程是一项重大且由我们共同面对的挑战。

广播媒体—社会发展的合作伙伴　　　　4

在危机中，为获得信息及信息处理优化模型，我们需要：

1. 在联合国教科文组织的协调下建立合作模式。
2. 建立允许并鼓励资源共享的国际平台。
3. 进一步开发"社区信息通讯中心"的作用和价值，将其作为核心单位贯穿于危机管理的各个阶段。

广播媒体—社会发展的合作伙伴　　　　5

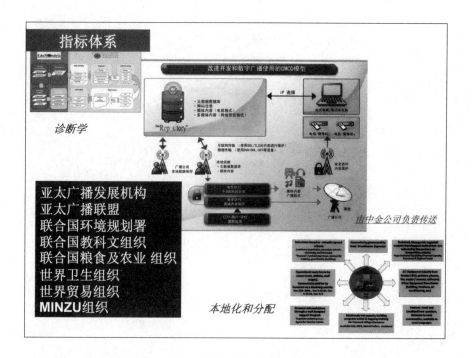

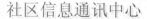

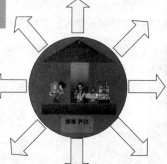

广播媒体提供的现有的
初期应用实例

点击链接： http://vimeo.com/16389768

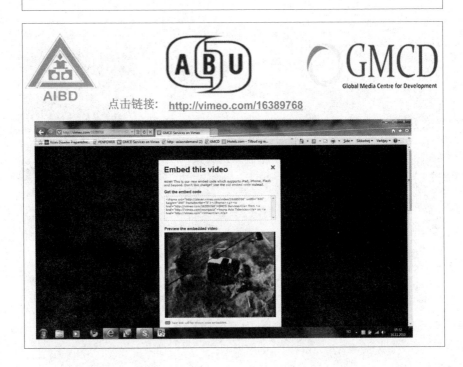

处理幼儿成长中挫折的成熟解决方案

幼儿园　　关于幼儿的信息　　通讯工具

 文本或
短信息

 图像

 实时视频

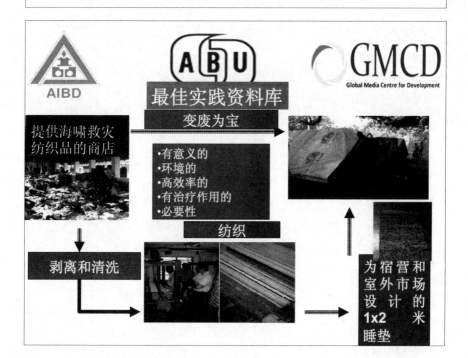

广播媒体已经就位
蓄势待发

四、专题会议

（生态安全：国际多边合作的战略构想）

全民参与，加强生态安全教育

柬埔寨王国政府国务部长　郭平

尊敬的主席，
尊敬的各位代表，
女士们、先生们：

下午好！

此次能有机会在如此盛大的会议上就生态安全——这一当今世界所面临的重大问题——发表浅薄的见解，我深感荣幸。众所周知，地球是人类所能生活居住的唯一家园。因此，地球居住条件的恶化也就意味着人类的生存在面临直接的威胁。然而，由于现代技术的发展和工业化的影响，当今世界正在承受这种威胁。可以说，人类是自食其果。

今天，我们相聚在这里，试图在人类遭遇到更多环境危机和灾难之前找到阻止环境急剧恶化的方法，否则，整个地球生态环境将会完全崩塌。正是人类的索取无度，才导致了我们赖以生存的环境日益恶化这一事实。当前日益明显的趋势是：由于工业化、商业化和物质化成为经济发展模式的主流，整个社会只关注如何赚钱，而人类的灵魂却丧失了理性，无法辨别"需要"和"想要"。　结果对资源的需求给我们的地球母亲造成了极大压力，导致了对自然生态的巨大破坏，由此威胁到了人类自身的生存。如果所谓的丰富的物质生活是以人类自身的生存为代价，那么这个代价真的是太昂贵了。

人类是地球上最具智慧的生物，近年来科学技术的发展充分证明了这一点。然而，如果人类只注重科技的发展而忽视了道德标准，忽视了对他人的关爱，对一切有感觉的生物以及对大自然的关爱，那么这样的进步并不一定能为人类带来安全感和幸福快乐。

尊敬的主席先生，要解决上述问题，我们需要从不同角度考虑，诸如科学、哲学以及自然规律。我们应该记起老子曾教导我们"人类要与自然和谐相处"，意思就是人类应当遵循自然规律。　此外，圣雄甘地也说过"虽然地球能够满足人类的需求，却填不满人类欲望的沟壑"。

在这个重大问题上，佛教典籍中有关贪婪、道德堕落及其对自然的影响的论述可以解释造成当今生态难题的原因。如果一个国家的国王或统治者缺乏道德观念，那么他的幕僚、各级官员、商人以及老百姓就不会遵守国家的法律秩序。此时，为了满足贪欲，人们就会无所不用其极，不惜违反自然规律（或者佛法）。　结果，太阳、月亮和风改变了往常的运动方式，造成了季节反常。当这种情况发生时，就是自然母亲或者说我们所称的上帝发怒了。于是，雨水不再像往常一样按时降临。在这种反常的季节和降雨条件下生长的庄稼无法满足人类健康的需求，因此，人类的寿命缩短了。

相反，如果人们遵守国家的法律秩序，尊重自然规律，并且具有良好的道德水准，自然母亲或者上帝就一定很满意。因此，人们就可以平和、幸福、富足和长久地生活。　这也就是佛祖所说的"种瓜得瓜，种豆得豆"。

此外，佛祖还指出所有罪恶都源自人们的贪婪、自私、仇恨和无知。虽然贪婪、自私、仇恨和无知是人类的天性，但是我们依然能够把握灵魂，战胜那些阻碍我们达到真正的幸福生活的恶念。

因此，佛教提倡与自然友好相处，以友爱和睦的眼光看待动植物与人类之间的关系。并且，我们应该记起佛祖生命中的三件大事：出生、启蒙和圆寂。所有这些都发生在成长在大自然里的树下，因此，佛祖教导弟子要在森林中进行冥思。因为欢乐平和的自然环境被认为可以促进灵魂的提升。

不仅如此，同情心一直都是佛教的道德准则。同情心不仅是指不伤害其他生灵的生命，还包括减轻他们的痛苦。

迄今为止，我们所遇到的生态问题源自人类前所未有的、无止境的欲望的膨胀和道德的堕落。解决环境危机的核心在于控制和消除人类灵魂中的贪欲。为此，我们应该践行仁爱与同情，培养勤俭节约的习惯，保持对生态安全负责的态度。我们需要共同努力来降低消耗，节约能源，避免对空气和水资源的污染，自觉自愿地为动植物的生长留出时间和空间。

我们作为领导者的职责就是教育我们的公民——当今和今后的人们——提升和发展道德水准，防止道德的进一步堕落。我相信我们能够在这条特别的精神之路上携手共进，重建我们的自然环境。如果我们善对大自然，大自然也同样会善待我们。

地球不会再有另外一个，因此，让我们每一个人都珍惜我们的星球、我们的家园吧！

祝愿每一个人都永远幸福快乐！

谢谢关注！

携手推进国际生态安全合作

塞舌尔共和国驻华特命全权大使　菲利浦·勒加尔

尊敬的主席先生，

各位嘉宾，

女士们、先生们：

下午好！

首先请允许我向柬埔寨王国政府的热情接待和周到安排表达谢意，我们此次金边之行成果丰硕，而且愉快顺利。

我还想特别向国际生态安全合作组织主席蒋明君先生致以谢意，感谢他邀请我参加此次大会并做主旨发言。

主席先生，今天上午，我国外交部长发表了关于塞舌尔应对气候变化和生态安全积极立场的演讲，并谈及了关于海洋可持续发展的一系列挑战。就此，我想感谢世界生态安全大会组委会就这个重要而紧急的议题给海岛国家，特别是小岛发展中国家提供机会阐述立场。

主席先生，此次会议的议题使我们有幸不再以划分领土界限的方式来讨论国际生态安全问题。更准确地说是，各方开展国际多边合作，从全球战略的高度迈进国际生态安全的远景目标。

然而，许多工作还有待完成。对政治家和政策决策者而言，就战略规划达成一致，当然绝非易事。一方面因为在哥本哈根气候变化会议及之后相同议题会议的讨论成果当中，可供他们考虑的一致性意见实在有限。另一方面，因为研究气候变化的政府间专家组发表的争论和文章中带给公众的意见和假设分歧过多以至于令人费解；而类似美国哈特兰德研究所、麻省理工学院等研究机构发布的报告，一旦归纳成新闻消息公之于众，又经常被全球公众理解成世界末日或过于乐观，形成两种极端。

但问题是，如果没有国际生态安全多边合作战略规划，针对威胁各国和地区经济的全球环境恶化问题，任何寻求解决之道的努力都将难有作为。

这一点对小岛发展中国家就更是如此。

主席先生，从不足500平方公里的陆地面积和约8.5万的人口来看，塞舌尔是个小国。如果我没搞错的话，这一人口大概相当于生活在加利福尼亚一州的美国籍柬埔寨人或者暹粒市人口的一半。

但当您从130万平方公里的海洋专属经济区这个角度考虑时，塞舌尔又是一个大国。实际上，塞舌尔海洋专属经济区面积位列全球第24，在南非和毛里求斯之间。这就解释了为什么我的国家被比作大水族馆里的一条小鱼。

然而，在应对气候变化带来的挑战和威胁时，我们不能仅仅依靠自身有限的人力资源和自然资源。

我们需要彼此协作。

人类作为独特的社会群体，确实需要彼此协作。不可能作为孤立的个体来谋求生存和繁荣。国家亦是如此。国家，作为社会实体，在现今相互依存的世界里，彼此间的依赖程度日益加深。因为在很多上述提及的挑战和威胁的解决过程中，位列最先和最重要的决议环节都是次区域性和区域性层面的。

在这一背景下，国际多边合作搭建了实用而及时的平台，以通过磋商与对话达成各方一致的决议。战略性的远景规划方可产生。

主席先生，今天，我想强调三种国际多方合作渠道，通过它们，我们需要针对整合协调一致的政策和行动计划付出更多努力。

第一种渠道即制度化。

我所指的是地区性政治和经济合作伙伴关系。同时作为驻中国、日本、韩国和越南的大使，我想强调我们与这些国家开展合作的重要性，特别是与非洲合作的重要。而且，履行合作的各成员国不应把精力仅局限于扶贫、健康、基础设施及食品安全，而应当把生态安全作为一项优先考虑的内容增加到合作中。这一点是重中之重。

在上述合作以及地区性和次区域性组织间的合作中，很多工作都为联合国和国际论坛中产生实质性结果做了铺垫。据此，在这一层面的讨论将需要我们长久的投入，并需要我们就在国际组织间推广可持续发展的最佳原则和做法达成一致。

主席先生，第二种和第三种渠道在于推广顺畅的多边合作。

第二种渠道即借助世博会和国际博览会。

在过去两年间，我担任塞舌尔上海世博会总代表开展工作，并担任上海世博会指导委员会委员。

如您所知，上海世博会的主题是"城市让生活更美好"。在塞舌尔，截至目前，我们只有一座城市，即维多利亚，人口仅3万，是世界上最小的首都。若把塞舌尔作为一个模型，向其他参加世博会的190个国家和56个组织以及7300万的参观者宣传我们的城市规划和城市发展战略，未免有些夜郎自大。

但是，我们想要向参观者传达和分享的是一条非常简单的信息，即塞舌尔同自然独特的关系：我们相信，自然不属于我们，而我们归属于自然；及我们如何将这种传统信念转化为国家政策，从而达到经济与社会发展、自然遗产保护和文化遗产传承的平衡。

我们希望我们的展馆尽可能地源于自然，成为一个生态友好的警世钟，如果这样说并不为过的话，我们的目的达到了。塞舌尔馆所在的非洲联合馆的总参观人数达到了2200万。2200万！我不敢说参观者们都收到了这一信息，但是，我肯定其中一大部分人收到了。

在2012年，丽水世博会以"生机勃勃的海洋及海岸"为主题，也将如期而至。我相信所有出席今天会议的海岛国家和有漫长海岸线的国家，也包括国际生态安全组织都将会把参加丽水世博会看做推动我们"更美好的星球"这一共同议程的必需。

主席先生，总结来讲，我想说国际生态安全合作组织是顺畅国际多边合作的第三种重要渠道。它就气候变化这一迟早会攸关各国的议题，提高了国际社会的意识和理解，并发挥了鉴别和促进科学研究及经济计划的催化作用。这一切使我们更充分地准备好适应变化中的世界、我们面临的经济和社会领域的各种风险，以及这些风险中孕育的机会。

谨此特别向国际生态安全组织开放协作的精神及对塞舌尔的特别关注致以敬意。

谢谢大家！

堪萨斯市与国际生态安全合作组织的共同目标

美国密苏里州堪萨斯市首席环境官　丹尼斯·墨菲

首届生态安全大会参会者，各位嘉宾：

感谢总干事蒋明君邀请我参加首届世界生态安全大会。作为堪萨斯市的首席环境官，我很荣幸能代表我的城市讲话。

密苏里州的堪萨斯市位于美国中部，一个我们通常称做美国心脏的地方。它位于美国最大的河流之一——密苏里河的两岸，大约离密苏里河与密西西比河交汇处200英里远。在18世纪，堪萨斯市是探险家、早期开拓者和商人们前往大平原、落基山脉和美国西部沙漠地区的出发地。

在今年10月举行的第14届埃德加·斯诺研讨会上，堪萨斯市很荣幸地请到了总干事蒋明君和其他中国高级官员。研讨会在中国北京和堪萨斯市轮流举行。今年的研讨会以清洁能源和绿色解决方案峰会为主题。在这次研讨会上，总干事蒋明君就生态安全观做了精彩演讲。

尽管"生态安全"这个名词对堪萨斯市来说是全新的，我们对这个观念和生态安全的要素却很熟悉，它代表了我们和国际生态安全合作组织的共同的目标。在过去的几年里，在市长马克·芬克豪斯和市议会12名成员的领导下，气候变化和可持续发展已成为我们市政府工作的重中之重。在堪萨斯市，"可持续发展"意味着我们做出的决定和采取的行动能同时促进经济活力，提高环境质量和促进社区的社会平等。

在清洁能源和绿色解决方案峰会上的演讲中，总干事蒋明君提到，由全球变暖和人类经济活动引起的生态灾难对人类生存和发展构成了严重威胁，灾难的频率、伤亡人数和财产损失仍在不断增加。

我完全同意他的结论，并和他有同样强烈的忧虑。和美国其他城市一样，堪萨斯市正在与商业社区、忧虑的市民和都市区的其他城市一起，在市政运行和社区各个方面减少温室气体排放并共同应对气候变化。这样的城市数目在不断增加。

通过代表全美所有市的组织——美国市长会议，一个由1000多位市长签署的气候保护协议诞生了。这些市长致力于各自社区的温室气体评估与减排的工作。

芬克豪斯市长是美国市长会议气候保护协议的签署者之一。在2008年7月，芬克豪斯市长和堪萨斯市议会一致地采纳了密苏里州堪萨斯市气候保护计划。这个计划是由代表街区、忧虑的市民、环境组织、商业部门、地方能源公司和我们的都市区规划组织的社区领导们共同制定的。

堪萨斯的气候保护计划规定，到2020年堪萨斯市温室气体排放比2000年降低30%。这项计划包括55项具体的温室气体减排措施，致力于通过改善建筑、路灯、交通信号灯和水/废水处理设施的能源效率达到减排目标。它仍要求更多地利用低碳或无碳资源发电，比如风力涡轮机和太阳电池板。我们计划里的其他温室气体减排措施包括通过扩大公共交通和增加步行和自行车的使用减少交通系统中的矿物燃料的消耗。我们的长期目标是到2050年温室气体排放比2000年降低80%。

这些目标是宏大的。但是气候变化对国家安全和人类生存的威胁是如此强大，除非我们达到这些雄心勃勃的有关温室气体减排的目标。在美国，我们已经认识到，由于缺乏足够的国家层次的推进，地方政府必须采取果断的有意义的行动，才能逐渐减轻不利的气候变化，并削弱它的影响。地方政府仍在继续要求国家领导人制定温室气体减排的措施，与国际社会的所有成员共同合作解决气候变化的问题。

因此在堪萨斯市，我们相信国际生态安全合作组织为实现联合国千年发展目标——消除贫困和饥饿、保证环境可持续性和发展全球合作伙伴关系——所做的工作至关重要。我们赞同国际生态安全合作组织的努力，并为首届世界生态安全大会的召开表示祝贺。

说到全球合作伙伴，我现在与美国一家叫做可持续发展研究所（ISC）的非营利组织一起工作。作为一名ISC中国气候变化领导学院的教员，我希望在接下来的两年里能与中国省市、社区、非政府机构、大学合作，一起制定与实施气候保护策略。这样的合作伙伴关系是我们国家的地方政府资源共享和共同工作的重要方式，地方政府共同而非单独地解决我们所面临的最具挑战性的全球环境问题。对气候变化的有效反应不是一个国家就能做到的——它需要世界各国共同努力的累积影响。世界各国的地方政府都在自告奋勇地提供这一领域所需要的领导阶层。

能在今天与你们分享我们在堪萨斯市举行的一些活动，并从你们这里学到如何将生态安全理论与实践运用到我们市政府的运行中去以利于我们更好地开展工作，我感到很荣幸。我们拥有共同的目标——协调我们的政策和行动以促进全球生态安全并为我们的子孙留下积极的遗产。

　　我代表芬克豪斯市长、市议会和堪萨斯市的所有居民，感谢你们邀请我参加这次重要会议，为我们提供了一个与国际生态安全合作组织对话的机会。我们希望这次对话能带来与你们合作的其他方式，为我们的世界和其他民族创造一个更好的未来。

重视生态同盟的活动方向与课题研究

国际生态生命安全科学院副院长　马拉杨

尊敬的女士们、先生们，各位同仁！

请允许我代表国际生态生命安全科学院向各位亲爱的参会代表与朋友们表示衷心的问候，感谢你们在这片热情好客的土地上组织并举办首届世界生态安全大会。

我们能够聚集在这里，那要感谢我们非常尊敬的蒋明君先生以及他的团队的巨大努力。当然，也向柬埔寨政府表达我们深切的致意与感谢。

我的发言想引用18世纪末期伟大的法国学者拉马凯的话作为开头："可以这样说，人类的意义在于毁灭自己的种族，把地球变成不适合居住的地方。"

遥远的过去，他已经预示并定义当时还无法想象的当今社会危机——技术文明危机，人类与自然复杂关系的危机。20世纪人类还呼吸着新鲜的空气，饮用干净的水源，世界似乎辽阔无边，自然资源无穷无尽。几十年过去了，世界已经处于人类制造的、可怕的生态危机的边缘。如果人类仍将继续这条道路，那就无法逃避再过几代人之后的灭亡。很多生态学家和悲观主义者都持这种观点，或许他们是现实主义者。

我们称做的生态危机是由什么引起的，为什么它会在20世纪末产生并在当今社会发展？一些学者认为，这是基于两个原则的基础之上：人口数量的增长（20世纪初全球人数约10亿，21世纪初人类总数增加了6倍）和科技革命，大大促进了自然资源的消耗（能源消耗增长了10倍，资源使用量增加了9倍，比人类增长的速度还要快）。

"生态危机"的概念在1972年第一次被提出，在研究全球现代问题的罗马俱乐部发言稿中出现。梅多斯领导下的团体，在其论文中，作出以下的结论：根据经济发展趋势和节奏的提高，人类将在2100年走向灭亡。那时大多数人类将死于饥饿与资源匮乏。自然资源不能满足人类基本生存生产原料，由于环境污染地球成为不适合人类居住的地区。

综上所述，人类目标（提高生活水准）与自然条件可能性的矛盾已经摆在眼前。人类对自然的破坏与干预将导致自然的毁灭，也就是说人类的毁灭。

感谢各国有先进思想的人士的理智与坚持。1992年7月，里约热内卢召开了联合国环境与发展会议，会议产生了名为《21世纪议程》的全球活动计划纲领性文件。在世界历史上首次通过地球文明发展国际战略方针，称为"可持续发展"。虽然一些学者认为需要使用另一个概述——人类生存纲领。因为不断被掏空的自然资源不能作为经济增长的支柱。以下是我们生态同盟——国际生态生命安全科学院，国际生态安全合作组织以及其他权威性的、关心生态安全的国际组织研究所作出的20世纪趋势，揭示可能的活动方向与研究课题。

这些趋势根据重要性用以下顺序展示危机问题：

1．全球气候变化；

2．人类机体的损坏；

3．水的自由使用与质量；

4．矿产能源资源的枯竭；

5．生物多样性的消失；

6．臭氧层的破坏；

7．土地利用的特性。

我希望，在以后的发言中，这个名单会更准确、广泛，使我们能集中精力与资源对上述以及未列入的其他问题进行研究，并取得某些成果与答案。

我认为，我的同事阿巴隆斯基将会更准确地描述我们可以共同努力、解决为人类造福的问题的合作领域，并在国际公约与其他文件的框架下共同完成。

今天我们参加颁奖仪式，为那些在生态领域作出重要贡献的政治家、学者、企业家颁奖。我要对获奖者表示热烈祝贺与真挚祝愿。

只有在共同努力下，我们才可以保护地球免受生态灾难的侵害！

我想引用杰出的俄罗斯学者、哲学家尼基塔·尼古拉耶维奇·马伊谢夫最后一本著作中的思想，他提出把社会称做合理、有组织的，如果这个社会有能力实现与自然的和谐共处，协调本身发展的逻辑与自然逻辑。但马伊谢夫也把这种社会的存在称为乌托邦式。"这有可能吗？"马伊谢夫提出这个问题。自己也同时回答道："总是无法找到答案：行还是不行。"在书中他用这样的话结束："我们已没有时间，形成保障我们共同未来的道德基础……"

谢谢你们的耐心与关注！

国际生态安全课题研究与前景

国际生态生命安全科学院副院长　　斯坦尼斯拉夫·阿巴隆斯基

尊敬的女士们、先生们，亲爱的嘉宾：

我非常高兴代表国际生态生命安全科学院在首届世界生态安全大会上向各位致意。

我知道，大会的组织方经历并克服了怎样的困难：我要向蒋明君和他的工作人员们，以及所有参加大会筹备组织的柬埔寨政府代表，致以深切的感谢并鞠躬。

我希望，这个创举将使更多的人，像这次大会的参与者一样，关注人类面前的生态安全问题。联合国把生态安全列为工作中一个重要课题并不是偶然。

必须指出，在国际生态生命安全科学院成立之日起，就把生态安全当做最基本的国际工作来开展。

科学院主席团和主席团主席鲁萨克·奥列格，深度关注生态安全领域活动，并坚持国际生态生命安全科学院积极参与并与国际权威组织合作的工作方针。

在经过巨大的组织与科研工作后，从2000年开始，国际生态生命安全科学院在联合国新闻署注册，2003年起国际生态生命安全科学院成为联合国经社理事会成员单位。

我们做了大量的、繁多的筹备与科技工作，使科学院进入联合国新闻署，并参与其中的工作。但科学院进入联合国机构的最基本目标为：第一，使国际生态生命安全科学院的学者与专家得到接触世界科学成就信息的可能性，来促进科学水准的提高；第二，把国际生态生命安全科学院在生态与生命保障所有领域的新技术成就，在全球各种组织机构的帮助下，让更多人扩大信息视野；第三，在联合国与众多基金的支持下，参与更多国际计划与项目。

2004—2010年当中，国际生态生命安全科学院成员的工作，与下列联合国开展的计划完全符合：

——自然保护活动；

——气候变化框架协议；

——发展目标下保护并合理运用资源；

——大气层的保护；

——切尔诺贝利灾难问题与核能源发展。

在自然保护项目研究上，科学院的学者作出了很大的贡献，我们的专家首次提出的生态科技方法基本标准的划分成为通用的方法划分。此方法最重要的、原则性的理论为生态资源与生产科技过程生态划分的概念——系统分析基础上的生态活动系数。

科学院专家研究其他的一些重要问题，如：

——无控制的电磁流不断增长，以及其对空气的污染；

——工业生产所产生垃圾的污染 （如戴欧辛）等。

根据一些评估，这些污染对地球上的人类、植物与动物群具有无法估算的危险，放射污染也同样极其危险。科学院的专家们已积累了大量理论与实践经验，使全球各国受益。

解决地球生态安全问题的唯一途径，是世界各国的合作：发达国家与发展中国家的合作。所以我们全力支持蒋明君创建一个解决生态安全问题、保护生态环境、实现可持续发展战略的机构——国际生态安全合作组织。并在不久之后创建了国际生态生命安全科学院中国分部，选举蒋明君为中国分部领导人和科学院副院长。

实践证明，我们的选择完全正确。国际生态生命安全科学院中国分部在改善国家生态安全方面取得很多重要的成果。这些成果，在国际生态安全合作组织出版的《国际生态与安全》杂志中大为推广宣传，取得广泛的知名度。

国际生态生命安全科学院主席团高度评价中国同仁所做出的工作，并希望将来继续巩固与发展合作形式。

在我们看来，具有广阔前景的有以下领域的项目：

1. 为未研究透彻的问题完善科学理论基础，使我们能采纳行之有效的决策促进可持续发展：能源（能源有效利用与能源使用）、交通、工业、海洋资源与土地利用等。

这个领域的基本与终极目标为能源产业减少对大气的不良活动，项目针对拓展可再生与可循环能源系统，特别是利用非传统与可循环能源（太阳能、潮汐能、风力等）。使用减轻污染与更高效的生产手段，输送、分配、利用能源。这个目标必须反映出必要的公正性与那些发展中国家必需的与不断增长的能源需求，结合那些国家财政收入主要来源于工业生产、加工与出口，或者使用传统煤炭、石油作为燃料的国家的实际情况，这些国家在向清洁能源转型过程中的困难，还有那些被有害活动引发气候变化的国家的薄弱性。

2．防止臭氧平流层被破坏。

3．大气层跨境污染，包括：对放射与电磁流污染的监控与采取预防措施；研究放射与电磁活动对环境指标变化的影响，并研发稳定方法；研究工厂排放的重金属氧化物对空气的污染；工业垃圾对地球表层的污染。

4．防止对地球进行抢掠性森林砍伐，这种砍伐将对人类正常生活导致灾难性后果。

必须指出，对保障生态安全项目全球层次上更深层的研究，能让我们更详尽地勾勒出利用科技新成就，在紧急时刻保障生命安全与国民健康的道路。

在当今社会，安全问题的最主要方向为科技所引起的灾害危险，根据联合国相关部门统计，技术灾害占紧急情况总数的70%以上。

国际生态生命安全科学院专家所从事的重要工作方向，包括：

——规划与合理使用自然资源的综合措施（土地、水、能源、生态资源等）；

——促进可持续的农村经济和农村的发展；

——生物工程技术的生态安全运用；

——维护水系统，保护、合理利用与开发其生命资源；

——淡水资源质量的保护与补给：运用综合手段开发水资源，深入研究水经济与水利用；

——对化学有害物质使用的生态安全管理，包括防止有毒危险产品非法国际运转；清除有害垃圾；防止化学有害垃圾在国际上非法转运；清除固体垃圾以及净化废水的问题；清除放射型垃圾；

——与森林砍伐、土壤沙漠化作斗争；

——推广生态无害技术，在所有活动方向合作并建立科学力量；

——对环境保护问题与解决过程中数据的统计；

——建立生态与经济因素综合统计系统。

国际生态生命安全科学院的专家与学者们在以上领域较高的水准让科学院更有信心在未来国际活动的拓展，在总部与各个地区分部领导积极倡议的基础上，科学院成员利用自己的研究成果，实践并参与到国际运作当中，包括联合国与许多国际基金项目。

科学院主席团也极其关注全球专家评审委员会的工作，此委员会由科学院高层次的专家组成，对所有重大的项目设计进行评估保障。我们希望，将会有更多的不同国家的，关注生态安全问题的专家加入我们的委员会。

今天我们在这里举行表彰授奖活动，一批在生态安全领域作出重要贡献的政治家、专家、学者和企业家人士将受到表彰。我希望这样的表彰，是我们在新的高度上共同致力于推进生态安全与可持续发展工作的一个新的起点，面对当前频发的自然灾害与生态危机，我们所做的远远不够，我们要做的事情还有很多。

最后，我必须指出，仅一次活动，甚至像首届世界生态安全大会这样的重大活动，无法解决人类所面临的生态安全领域的所有问题，而是需要世界人民与国家政府的共同努力，所以，等待我们的是每日细致复杂的解决生态安全的工作。而且我们不能忽视，工业生产的发展总是带来生态问题的增长，这些问题亟须解决。

生态安全：强化非政府组织的广度和深度

联合国环境规划署俄罗斯委员会协调官　根纳吉•施拉普诺夫

尊敬的主席，女士们、先生们！

　　我很荣幸能够在这里，这个全亚洲最美的首都之一——金边，参加并在尊敬的各位面前发表演讲。我仔细倾听了同仁们的发言，并获得无上享受。这些发言无论在理论或者实践上，都有高度丰富的内容。我现在简短地叙述一下自己对讨论议题的看法。

　　早在1996年，俄罗斯海参崴举行的一次重大国际论坛上，蒋明君先生就提出以下主张：生态安全、资源安全、可持续发展将占据重要的地位。

　　经过15年，现在可以坚定地说，这个预警已经在某些领域成为现实，这些问题也在国家外交战略占据了越来越巩固的位置。从事这些问题研究的政治家、学者、社会活动家在全球各界也更加具有影响力。

　　生态安全——国家安全的重要组成部分。

　　俄罗斯作为世界大国之一，需要和其他国家一起在生态安全领域承担自己的责任。

　　联合国环境规划署俄罗斯国家委员会积极地与国际生态安全合作组织、国际生态生命安全科学院、国际商业技术大学等国际组织和学术机构合作，进行多边、跨行业的生态安全领域新理念、新技术交流。

　　我认为，通过非政府组织、学术机构的紧密合作，积极推动生态安全领域国际合作的广度和深度，可以对中央及地方政府的措施进行监督。我认为，这也是非政府组织的一个重要任务。

　　俄罗斯和中国有着大约4300公里的边境线，生态安全问题至关重要。

　　举个例子：界河阿穆尔，中俄之间的长度为3000公里。阿穆尔河受污染的问题在现实中非常严峻。

　　根据媒体的报道，仅中国黑龙江省就要向阿穆尔河流域排放几十亿立方米的未经处理的工业及生活废水，这要比俄联邦滨海区流域的指标高得多。双方都筹划在阿穆尔河上建设水电工程，水电工程的建设需要双方国家高层机构以及社会组织严肃认真的研究。

虽然目前俄中共同的河水检测不断进行，我们作为非政府组织必须促进两国《保护与利用界河合作公约》的签订过程。

另一个重要的国际合作领域是保护与维护世界自然遗产。

大家都知道，贝加尔湖——地球最深的湖泊，是世界最大的淡水储藏库，拥有独有的生态系统。

早在1994年9月，在乌兰乌德市举行的"贝加尔地区—世界发展模式区域"国际会议，在促进贝加尔湖研究中起到了重要的作用。

之后在世界银行的倡导下，贝加尔湖生态旅游发展总规划的科学研究得以进行。这个规划的基本目标是在对自然环境危害极小的情况下，得到最大的经济利益。

随后，世界银行投资的"贝加尔湖流域自然资源管理与生物多样性保护"项目策划完成。

项目在俄罗斯—美国联合项目（戴维斯项目）"俄罗斯贝加尔湖流域土地利用政策综合目录"的框架下，进行了实地调查工作。

欧洲社会针对贝加尔湖地区的塔西斯项目确定了十分重要的自然保护任务。此项目的一个基本优势在于：项目进行过程中给予了项目操作人员认识欧洲组织在不同方向，包括自然资源与环境质量管理领域中的实践工作经验。

这些简短的国际合作名册证明，贝加尔湖的保护工作在不断地完善，并拥有很广阔的前景。但是，在转向具体的自然保护项目操作领域中，对国外投资者却没有完全开放贝加尔湖生态系统现状的信息。

更加延缓项目进程的因素还有俄罗斯社团组织不够积极的态度，具体在贝加尔湖地区，是那些能够对自然利用与保护问题做出协调作用的社会组织。

贝加尔湖和它的生态系统具有全球意义，对它的保护是整个国际社会的责任。许多具有先进思想的学者都认同这个观点。我坚信，解决贝加尔湖问题上的国际合作将会不断拓宽发展。

大会的倡议与筹办方是国际生态安全合作组织。目前已有来自亚、欧、美、非洲的许多国家表现出极大的兴趣。

对在座的各位不需要阐述老年问题的必要性，我只想说明，这个问题也和生态问题紧密相连。

在座的嘉宾中有许多我的同行，已经开始积极地进行大会的实际筹备，这又一次证明，我们的合作将不断地拓展并一定会有丰盛的成果。

谢谢！

青年力量在全球可持续发展中的作用

联合国青年技术培训组织亚太地区主席   山俸苹

尊敬的各位嘉宾，女士们、先生们：

大家好！很高兴与诸位相聚在金边，并以联合国青年技术培训组织的名义发表我的主旨演讲。我的演讲主题是"青年力量在全球可持续发展中的作用"。

联合国早就已经认识到，青年男女的想象力、理想、精力对于他们所生活的社会可持续发展至关重要。基于这一认识，联合国各成员国于1965年签署了《在青年中促进各国人民之间和平、互尊和了解的理想的宣言》。1995年，在"国际青年年"十周年的庆典上，联合国要求国际社会在下一个千年继续对向青年提出的挑战做出快速反应，以此强调它对青年所作的承诺。为此，它采纳了一个国际性战略，即《到2000年及其后世界青年行动纲领》（《世界青年行动纲领》），以便更加有效地处理青年男女所面临的问题，增加他们参与社会的机遇。

《世界青年行动纲领》是一个具体的行动计划，它包括了教育、就业、饥饿、贫穷、环境、药物滥用、少年犯罪、休闲活动、女童及年轻妇女、青年充分和有效地参与社会生活和制定决策在内的十个优先领域。在每一优先领域，行动纲领审视挑战的性质，并提出行动建议。这十个优先领域相互关联，不可分割。例如，少年犯罪与药物滥用往往是未能充分接受教育、充分就业、充分参与社会的直接后果。

在上述背景下，印度及孟加拉等国发起创建了全球性国际组织——联合国青年技术培训组织，并于2001年在联合国大会特别会议获得批准认可。联合国青年技术培训组织总部设在印度的加尔各答，隶属于联合国人居署二局。基于《世界青年行动纲领》，该组织的宗旨是强调妇女应有的地位和发展，以及她们的生存环境，致力于使青年们获得行动起来的力量，能够创造自己现在和未来可持续发展、政治和经济上切实可行、保证社会公正的生存环境。

为了能引起人们对青年面临的问题和挑战更多的关注，联合国青年技术培训组织确定了广泛的、对青年生活至关重要的工作范围，主要包括消除贫困和边缘化，实现就业、健康、人权、教育和环境保护，预防艾滋病以及倡导并促进妇女在可持续发展和环境问题方面的基层工作。

众所周知，青年是推动城市发展的重要力量。他们既是发展的主要人力资源，又是社会变革、经济发展和技术革新的主要推动者。他们的想象力、理想、充沛的精力和憧憬对他们所在地区的可持续发展至关重要。

世界各地的青年生活在不同发展阶段和不同社会经济环境的国家中，他们都渴望充分参与社会生活。但是，变动中的世界社会、经济和政治局势，使许多青年充分参与社会生活的目标难以在许多国家实现。这就是执行《行动纲领》的重要意义所在。

所以，联合国青年技术培训组织将发展教育列为重中之重。近年来，尽管各国在普及基础教育和扫盲方面取得了明显进展，但文盲的人数继续增多。为了鼓励发展更适合于青年及其社会目前和今后需要的教育和培训制度，联合国青年技术培训组织协助发展中国家扩充了青年人获得高等或大学教育、从事研究或接受自营职业培训的机会。鉴于这些国家所面临的经济问题和这方面的国际援助不足，联合国青年技术培训组织与各国政府和其他非政府组织一起，协助发展中国家的青年人在发达国家和发展中国家接受各级教育和培训，以及在发展中国家间进行学术交流。

除了教育之外，就业也是联合国青年技术培训组织工作的一个重要方面。如今，到处都有青年失业和就业不足问题。青年人很难找到适合自己的职业，而且他们还遇到其他许多问题，包括文盲和培训不足，再加上经济不时放慢和经济趋势全面转变，因此情况更为严重。

为了解决这一问题，联合国青年技术培训组织和各国政府及其他非政府组织一起，制订或推动赠款计划，为发达国家和发展中国家青年经营的合作社建立模范合作社；对于特定的青年群体，给予特定的就业机会。协调各国政府酌情在指定用来增加青年就业机会的经费范围内，划拨资源用于支持青年妇女、残疾青年、退伍青年、移徙青年、难民青年、流离失所者、街头儿童和土著青年的方案；鼓励各国政府特别是发达国家政府，在一些因技术创新而迅速演变的领域为青年人创造就业机会等。

比起教育和就业，饥饿与贫穷是一个更为基础的问题。今天，全世界有10亿人民生活在令人无法接受的贫穷条件中，其中大多数在发展中国家，特别是在亚洲及太平洋、非洲、拉丁美洲和加勒比的低收入国家以及最不发达国家的农村地区。

为了解决饥饿和贫穷，联合国青年技术培训组织与各国政府一起，加强贫困落后的农村地区的教育和文化服务，以及其他的鼓励措施，使之对青年人具有更大的吸引力；积极推动执行面向青年人的试验性耕作方案，并将推广服务扩大到农业生产和销售方面，使其不断得到改进；同各地区的本地青年组织合作举办文化节活动，加强城市青年和农村青年之间的交流；鼓励和协助青年组织在农村地区举办大、小型会议，特别是争取农村人口，包括农村青年的合作；根据农村的经济需要和农村地区青年人对发展生产和实现粮食保障的需要进行培训。在这类培训中，联合国青年技术培训组织特别注重青年妇女、留在农村地区的青年、从城市回到农村地区的青年、残疾青年、难民青年和移徙青年、流离失所者和街头儿童、土著青年和退伍青年，以及居住在冲突已经解决的地区的青年的技能培训和能力提升。

目前，联合国青年技术培训组织正在着手建立一个总的行动机制，这一机制将不断地调整必要的人力、政治、经济和社会文化资源。但这一行动机制的实施，最终都要由各国政府负责，并获得私营部门和其他社会组织的支持。只要发动社会各方面的力量支持我们青年人的行动计划，我相信，他们必然在城市可持续发展中发挥更大的作用。

谢谢！

"DS-循环经济系列技术"的特点与优势

北京汉祥环境生物科学研究院首席科学家　史汉祥

各位来宾，女士们、先生们：

下午好！我叫史汉祥，来自中国北京汉祥环境生物科学研究院。在刚刚结束的联合国国际生态生命安全科学院表彰授奖仪式上，我被授予"高级研究员"称号，对此我感到十分荣幸和激动。非常感谢科学院院长对我的评价和认可，借这个机会，向在座的各位领导、专家、学者简单介绍近年来我和我的团队所致力研究的"DS-循环经济系列技术"，请予指正。

随着工业的迅速发展和人口急剧增长，大量燃烧煤炭、石油所产生的化学物质以废气和烟尘等形式排放到大气中，超过了大气环境的容许量，造成当今世界生态灾难频发，人类生存受到严重威胁，形势已相当严峻。尤其是这些废气和粉尘的大量排放所造成的工业污染更是成为全球大气污染的罪魁祸首。

目前，在处理烟气污染上，世界各国大都在脱硫技术上做文章，90%采用的是石灰石—石膏法脱硫，其弊端是石灰石在吸收二氧化硫后产生的脱硫产物，大部分无法再生利用，占用了大量土地堆放并形成二次污染。

"DS-循环经济系列技术"的核心就是从解决工业立体污染入手，利用钢铁、火电等行业的钢渣、粉煤灰等固废做二氧化硫吸收剂，经多相反应器（此系列技术中的设备）实现高效脱硫，并将脱硫副产物用于改造盐碱沙荒地。为解决二氧化硫大气污染、废渣等污染物的综合治理和盐碱沙荒地改造三项难题开辟了全新的道路。在"废弃物资源化、减量化和无害化"的同时，逐步向资源循环利用延伸，形成完整的循环经济产业链。

"DS-循环经济系列技术"具有很强的适应性，它可以用于所有产生含二氧化硫烟气的行业，如燃煤电厂、铁矿烧结、燃煤炉窑、有色金属冶炼、化工、建材等；在脱硫剂的选择上，可根据烟气脱硫对象及其条件，因地制宜选择采用有色金属冶炼炉渣、炼铁炉渣、炼钢炉渣、电石渣、盐碱土、金属氧化物（如氧化锌、氧化锰、氧化镁）等工业碱性物料为脱硫吸收剂，从而极大地降低了运行成本。

　　该技术的主要特点是：（1）利用工业炉渣包括炼铜炉渣、高炉水渣、钢渣、赤泥、白泥、电石渣、粉煤灰、氧化锌烟尘、高炉粉尘、循环利用水淬渣的水等作为脱硫吸收剂；（2）采用标准化、模块化和无喷嘴结构设计；（3）采用特殊耐高温高分子工程塑料及一次整体成型的工艺技术，具有不结垢、不堵塞、防腐蚀、抗磨损等优点，真正做到脱硫率高、投资省、运行费用低、安全且使用寿命长等优点。

　　而该技术最显著优势是：可将吸收二氧化硫后的终产物通过脱硫产物中的钙、铁离子及水合硅化物与土壤中钠、钾、镁离子进行离子交换变成可洗出离子，脱硫产物中的酸性物与土壤强碱弱酸盐反应生成中性不溶物和易洗出硫酸盐的机理改变，制成确保脱硫产物重金属含量符合土壤改良所使用的国家标准的土壤调理剂和盐碱地改良剂。另外，炉渣脱硫产物中还含有土壤中需要的硫、磷、铁等营养元素，用以改良盐渍化土壤和盐碱沙荒地，促进农业的整体发展，形成从工业延伸到农业的循环经济产业链，从而成功解决了所有脱硫技术都面临的脱硫副产物无法合理处置和消纳的世界难题。真正做到了尊重生态规律，维持生态平衡。

　　近年来这项循环经济技术在应用和不断的改进中已取得包含17项发明专利，5项实用新型专利，并在中国浙江省、河北省、天津市和越南等国内外十多个省份，数十家大中型有色、黑色冶金、化工、电力等企业得到成功运用。

　　我认为，对于治理环境污染，不能够走"先污染、后治理"的老路，要彻底变革以牺牲环境为代价的传统发展模式，按照中国文化的精髓，用辨证的思维来指导经济发展。作为地球的公民，要肩负起保护地球的责任和义务，为创造人类美好的明天而共同努力。

　　谢谢大家！祝首届世界生态安全大会取得圆满成功！

生态新能源"海力富"的特点与价值

中国城市污水治理专家 邓楚柏

尊敬的各位专家，女士们、先生们：

你们好！非常高兴能够来到美丽的金边，与来自世界各国的政要、专家、学者共同分享海力富生态能源成果。此时此刻，我的心情无比激动，感谢蒋明君副院长对我的推荐，感谢联合国国际生态生命安全科学院对我工作的认可。借此机会，我想简单介绍一下海力富生态新能源的发表过程和主要特点，希望在座的各位嘉宾提出宝贵建议。

一直以来，IT产业被认为是20世纪最有商业发展价值的产业，而新能源产业则被认为是21世纪最有商业发展前景的行业。"微生物捕获碳循环燃料•海力富"无疑是21世纪最有社会效益和商业效益的新能源。

英国著名气候专家克瑞斯•古达尔出版了一本名为《拯救地球的十项技术》的著作，称人类如果使用十种新技术，将可以避免地球出现毁灭性的气候灾难。该十项技术其中之一是"碳捕获技术"。克瑞斯•古达尔认为由于目前可再生资源的增长已无法满足全球的能源需求，因此，寻找一种能够有效捕获和存储由发电厂所产生的二氧化碳的方法，已成为我们人类所要面对的最重要的挑战之一，"碳捕获技术"将是人们期待的最理想的方法之一。尽管碳捕获技术研究进展缓慢，但各国政府已开始意识到这项研究的重要性，新技术也将应运而生。

作为一种高科技生态能源，我们认为，与传统的生物能源，以及正在引起全球关注的民用核能、风能不同，"海力富"是通过光合微生物等多种微生物群落的混合发酵和生化合成的一种具有高效安全的生态燃料。

第一，"海力富"的原材料是人们在生产和生活中所排泄的废弃物，只要有人群活动，就不可避免产生这种废弃物，我们目前正为如何处理好这些废弃物而大费脑筋。而"海力富"把污水转化成生物能源，一方面减少污染源，另一方面生产出新能源，是节能减排的最佳范例。

第二，"海力富"燃料是一种耗能最少的燃料。"海力富"是通过微生物捕获碳元素合成的碳循环燃料，其合成过程是微生物在常温常压下进行的，与目前工业上常用的把木纤维素和淀粉转变为醇类燃料等相比更为节约能源。因为把木纤维素和淀粉转变为醇类燃料需要经过糖化和醇化等合成过程，而这些合成制造过程需要一定的温度和压力。而"海力富"不仅是在常温、常压下合成，而且它的原料主要是工业和生活排泄物以及空气中的二氧化碳，所以与目前其他生物燃料相比，不论材料成本还是制造成本，"海力富"燃料都是最便宜、最安全的。

第三，"海力富"是最安全的燃料。"海力富"的闪点为29℃，在使用、运输和储存方面都非常安全。此外，由于我们在生产的过程中对有害菌体进行了灭菌消毒处理，并在合成过程中对一些可能在燃烧时能形成有害气体的重金属，例如铜等进行了螯合屏蔽处理，经过初步的测试，"海力富"在燃烧时不会产生有害气体，即使在自然混合表面燃烧的条件下也不会产生黑烟。

第四，"海力富"有利于二氧化碳温室气体的减排。目前任何燃料在燃烧时都将通过氧化燃料中的碳释放出热能，在各种燃料中的碳成分最终转变为二氧化碳释放到空气中。而"海力富"在合成过程中把空气中的二氧化碳固定下来，同时由于微生物的生化合成作用，促进了污水中的成分向含有氢的可燃功能基团方面转化，充分发挥了氢能的热值，这就意味着在燃烧"海力富"时相同的热值所释放的二氧化碳要比燃烧其他燃料少得多，因此使用"海力富"燃料将可以大幅度地减少空气中二氧化碳的含量。

从以上的四个方面比较，我们可以自豪地说，"海力富"是21世纪最为生态、最为安全的燃料。一旦投入使用，它的商业价值和社会价值将不可估量！2009年6月，我们已经在北京召开了新闻发布会，正式将此项技术推广使用，目前已进入成熟试用阶段，我希望此项技术能得到各国政府的重视，在全球范围内实现它的价值。

谢谢！

五、附件

附件一 [*]：

第六届亚洲政党国际会议确认函

2010年8月29日于马尼拉

第六届亚洲政党国际会议
组委会：

　　根据亚洲政党国际会议今年在中国云南省昆明市召开的第十三次常务会议发布的通告，我们同意并接受国际生态安全合作组织的提议，将首届世界生态安全大会纳入同期举行的第六届亚洲政党国际会议全体大会。这次大会将于2010年12月1日至4日在柬埔寨金边市召开，因为双方都迫切地关注贫困、气候变化、环境恶化问题。柬埔寨副总理及第六届亚洲政党国际会议组委会主席索安阁下也提出了这样的建议。

　　与此同时，我们还接受了国际生态安全合作组织的提议，将邀请亚洲政党国际会议成员前往暹粒参观世界奇观吴哥窟遗产保护区。

您真诚的朋友
何塞•德贝内西亚
亚洲政党国际会议创始主席

---

　　* 本附件内容出自同页的文件扫描件。

附件二 *：

柬埔寨王国政府确认函

柬埔寨人民党中央委员会 2010年9月9日于金边

尊敬的蒋明君博士：

根据亚洲政党国际会议创始主席以及亚洲政党国际会议联合主席何塞•德贝内西亚2010年8月29日的来信，我很荣幸地向您确认，第六届亚洲政党国际会议全体大会将于12月1日至4日在柬埔寨金边市召开，首届世界生态安全大会也在此期间召开。为了供您参考，下面附上第六届亚洲政党国际会议全体大会的暂定议程，您会看到，世界生态安全大会开幕式在12月3日（星期五）的上午举行。

借此机会，我代表第六届亚洲政党国际会议全体大会组委会，邀请世界生态安全大会的与会者参加我们政党全体大会以及会外活动。

对此，我建议您以及您的代表们及时和第六届亚洲政党国际会议全体大会组委会协调议程，协议及安排大会的人员名单和需要的预算。

请阁下您接受我最崇高的敬意！

索安
第六届亚洲政党国际会议组委会主席
柬埔寨王国政府副总理
柬埔寨人民党常务委员会常委

---

* 本附件内容出自同页的文件扫描件。

附件三 [*]：

联合国大学支持函

2010年11月5日于东京

首届世界生态安全大会

组委会：

在首届世界生态安全大会即将召开之际，我谨代表联合国大学，向大会的顺利召开表示祝贺！

进入21世纪，各种自然灾害与突发性生态灾难越来越频繁，人类社会正面临全球气候变暖、土地荒漠化、水资源匮乏和污染、森林植被破坏、生物失去多样性、食品安全、粮食安全等生态危机的严重挑战，生态安全已经上升为影响全球发展的一个时代主题。

首届世界生态安全大会，为来自世界各国的政府、政党、议会，以及非政府组织、金融机构、企业集团及专家学者提供了一个分享经验、寻找出路的平台，有利于开辟不同层面的社会实践与紧密合作，建立全球化的生态安全合作机制。

---

[*]　本附件内容出自同页的文件扫描件。

　　作为联合国整体的智囊机构，联合国大学成立以来，一直致力于促进国际学术交流合作，开展以问题为导向、多种学科的全球事务研究，帮助解决联合国和会员国关注的有关人类生存、发展和福利等全球问题。联合国大学愿以积极的方式，参与到首届世界生态安全大会的各项工作之中，并期待首届世界生态安全大会取得丰硕成果。

　　预祝本次会议取得圆满成功！

武内和彦
联合国大学副校长

附件四 [*]：

中国国际问题研究基金会支持函

世界生态安全大会组委会

秘书处：

　　我会是中国外交部主管的全国性社团法人机构，其成员由中国资深外交官、著名国际问题专家、学者组成。其宗旨是推动学术界、经济界对重大国际问题进行前瞻性、战略性研究，并开展学术交流与国际合作。

　　关于国际生态安全合作组织与柬埔寨王国皇家政府、亚洲政党国际会议于2010年12月1—4日在柬埔寨主办《首届世界生态安全大会》的函已收悉，鉴于我会与国际生态安全合作组织是战略合作机构，我会同意协办《首届世界生态安全大会》。

　　特此函告！

中国国际问题研究基金会

2010年10月15日

---

　　* 本附件内容出自同页的文件扫描件。

附件五：

亚洲政党国际会议（金边）宣言

第六届亚洲政党国际会议全体大会正式通过
2010年12月1—5日于柬埔寨·金边

我们，来自阿富汗、亚美尼亚、澳大利亚、阿塞拜疆、巴林、孟加拉、不丹、柬埔寨、中国、朝鲜、东帝汶、印度、印度尼西亚、伊朗、伊拉克、以色列、日本、哈萨克斯坦、吉尔吉斯斯坦、老挝、黎巴嫩、马来西亚、马尔代夫、蒙古、尼泊尔、巴基斯坦、巴布亚新几内亚、菲律宾、韩国、俄罗斯、所罗门群岛、斯里兰卡、叙利亚、泰国、汤加和越南的89个政党代表团和政党领袖以及政府代表团为了第六届亚洲政党国际会议聚集到柬埔寨的首都金边。以"亚洲：追求一个更美好的明天"为主题的第六届亚洲政党国际会议全体大会与首届世界生态安全大会于2010年12月1—4日在柬埔寨金边市召开，大会由柬埔寨人民党与奉辛比克党、国际生态安全合作组织联合主办。

值得注意的是，这次大会的召开不仅是在亚洲政党国际会议成立10周年纪念日期间，而且也是"柬埔寨救国民族团结阵线"成立32周年纪念。我们发布《亚洲政党国际会议（金边）宣言》如下：

亚洲政党国际会议是为政党服务的公开论坛

我们重申亚洲政党国际会议是一个开放的、独一无二的为亚洲政党服务的论坛，并且成为党内对话与合作的支点，以努力实现亚洲繁荣和平发展为共同目标。亚洲政党国际会议引领着被普遍认为21世纪为亚洲世纪的道路。

我们满意和骄傲地注意到，在第一个十年，亚洲政党国际会议促进了亚洲各党派的交流合作，形成了竞争意识和形态；加强了国家和人民之间的相互理解与信任，并促进了区域合作。

我们每两年举办一次大会，2000年在马尼拉，2002年在曼谷，2004年在北京，2006年在首尔，2009年在阿斯塔纳，我们重申致力于亚洲政党国际会议宪章精神及亚洲政党国际会议宣言和宗旨。

我们重申要强调民主、人类安全、人权、尊严、自由、福利、健康、法律规则，各种族、文化、信仰共处的联合国宪章、国际法，和平共处五项原则和万隆原则的宗旨和精神。促进不同信仰之间的和谐共处是亚洲政党国际会议原则的基本信条。亚洲政党国际会议在充满活力的社会中展现着亚洲快速恢复能力的精神：成功对抗危机及以新观念来克服经济困难。

我们坚持的原则

我们发誓在不断成长的政治和经济多极化的背景下，要保证亚洲的和平、安全、稳定、繁荣，并坚持以下原则：

——国家的主权与领土完整；

——国家政治、经济、社会体系的自主权；

——互不侵犯，不干涉各自内政；

——和平解决领土纷争，并坚持国际条约与国际法；

——大规模破坏性武器的控制、解除、不扩散；

——拒绝各种极端主义、偏见和偏执行为。

我们特别支持在2010年条约审议大会上，缔约方一致通过的关于不扩散核武器的最后文件中指出的"所有国家都必须做出特别的努力，建立必要的框架，以实现和维持一个没有核武器的世界"，以及联合国秘书长提出的包括核武器公约谈判的五点主张。

我们支持常委会的倡议——与拉丁美洲和加勒比政党常设会议一起与我们非洲和其他洲的政党建立联系，通过对话和交流增强相互理解与合作，旨在最终召开一次全球性的政党大会。在亚洲政党国际会议常委会与拉丁美洲和加勒比政党常设会议协调机构的会议上重申了这个目标。

直视我们最大的威胁

我们意识到在我们这个时期，环境恶化以及一代传给一代的贫穷是人类面临的最大威胁。我们支持通过国际和国内的各种努力来减轻气候变化所带来的影响，以及缓解饥饿、愚昧和疾病。我们赞同联合国秘书长潘基文在加速实现联合国千年发展目标、促进国际协议以减少温室气体排放方面所做的工作。我们赞同亚洲政党国际会议在联合国大会中获得观察员地位的提议，以协调与联合国有关的项目和相关活动。

我们意识到经济控制了国家和区域发展的进程，亚洲的经济要改善其资金流动的管理和集中资源的能力。亚洲国家要通过平衡扩张经济及在全球经济增长中增加收益份额，寻找包容性的增长。可以通过消除发展鸿沟和通过全球化带来的连接性来加速经济整合。

我们呼吁区域性开放

我们意识到要通过区域性开放，即免除关税和非关税壁垒，加强合作及整合亚洲地区经济来提升地区间的贸易。我们需要与东南亚国家联盟和南亚区域合作协会的结构机制联系上；强化湄公—印度—中国经济走廊；将大湄公河次区域与南亚次区域的经济合作和亚太经济组织联系起来；还要将海湾合作委员会、中亚区域经济合作及其他区域合作机构框架纳入一个更大的泛亚工程。

在这方面，我们要提醒我们自己，我们的理想和最终目标是建立一个亚洲共同体，这将带来本地区所有人民共同的繁荣，并强调集体努力，特别是要通过政党的扩展以进一步加强所有国家和区域集团的协调与合作的重要性。

正如2010年7月亚洲政党国际会议在中国云南省昆明市举办的主题为消除贫困的专题会议上所倡议的，我们重申要设立一个亚洲反贫穷基金和一个亚洲小额贷款基金。我们建立的基金是一个全球性的反贫穷基金，我们的拉丁美洲和加勒比政党常设会议的伙伴以及非洲的政党将为我们提供咨询意见。

我们意识到，随着亚洲经济的增长，亚洲巨大的人口数量将对全球的粮食需求施加更多的压力。我们现在必须建立一个可持续的农业增长中心。湄公河流域可以成为这个中心，但也不限于东南亚的资源。我们敦促湄公河流域的国家保护好各自的水资源，发展农业，特别是有机产品，不仅要减少人民的贫穷，还要维护地区的食品安全，抑制全球食物价格的通货膨胀。

应对气候变化与生态危机

我们认识到，要改变传统的经济观念，以便承担共同的责任来应对气候变化所带来的危险。目前，我们的国家必须应对更多的生态灾难，例如超级台风、特大洪水、特大泥石流。我们必须重新审视传统的能源文化，并且推动可再生和清洁能源的创新应用。

我们必须通过把工业变成环境的朋友而不是敌人来加强人类与地球生存的纽带。公共政策必须鼓励"绿色企业"的兴起，并利用适应环境要求的技术。我们必须铭记我们不能对地球家园为所欲为，我们仅仅是它的管理者，我们的活动必须为了下一代。

我们将迎接可再生能源

我们认识到发展可再生能源不只是为了减少二氧化碳的排放，也是创造经济活动与可持续发展领域的一个机会，提高大众收入和制造就业岗位。对亚洲的可再生能源的明智开发，充分利用其多元化和广泛可用性，成为我们亚洲能源战略的最终目标。

因为要完全开发可再生能源和绿色技术需要一定时间，我们必须，至少在可预见的未来，要继续依赖传统能源来作为亚洲发展的燃料。因此在一个油价震荡的时代，我们也必须主张在亚洲很多地区，既包括对内陆也包括对近海地区开发石油和天然气的储备。

我们要保护生物多样性

我们认识到生物多样性在文明社会的发展中扮演着关键的角色。人类必须积极热心地保护这片广袤无垠的物种多样的土地。我们呼吁成立一个"亚洲生态安全研究中心"来支持亚洲的可持续发展并鼓励在可持续发展这方面的先驱精神和全面革新，特别是在畜牧业、林业和渔业方面。

积极应对自然灾害

我们也呼吁亚洲国家紧密合作应对经常发生的自然灾害和突发性生态灾难，例如地震、海啸和洪水。我们必须竭尽全力预防这些灾难，并提供救援和重建工作。我们敦促亚洲政党国际会议医疗救援论坛的召开，正如2010年5月在马来西亚兰卡威召开的第一次亚洲政党国际会议医疗救援论坛督导委员会所批准的，以应对自然灾难带来的威胁。我们赞成2011年5月在马来西亚召开主题为"自然灾害与环境保护"国际会议的决定。我们同意《世界生态安全大会（吴哥）议定书》，并支持国际生态安全合作组织的各项发展目标，建议各国按照本国宪法和政党纲领来应对气候变化和生态环境的日趋恶化。

青年和妇女组织的作用

我们认识到青年和妇女获得并控制资源、取得机会深入到生活各领域对可持续发展和整体的繁荣很重要。我们认识到没有他们的积极参与，社会文化和现代化的进程是无法完成的。

推动政党谈判解决争端

我们在亚洲各冲突地区寻找和解的途径。我们的最终目标之一是代表全球和平进程来推动世界政党团体；这个地区的经济和政治一体化；经济秩序的稳定；债务免除，采取措施应对贫穷、腐败和气候变化。我们亚洲和世界各政党团结在一起并采取统一行动，可以在消除东西、南北鸿沟方面献出自己的力量；预防冲突与威胁；从恶化的生态环境中拯救我们的共同家园。

保护亚洲与世界遗产

我们必须强调，保护亚洲辉煌的世界遗产、其文化和价值的重要性，在不侵犯个人和集体权利的程度上，在全球化的背景下继续指导和鼓舞我们现代化的进程。我们呼吁建立一个机构来为可持续发展，特别是在农业、环境、能源、和平和文化领域作出杰出贡献的人士授予大奖。

祝贺柬埔寨皇家政府和国际协调委员会在吴哥世界遗产保护方面的成就，18个国家、25个机构和60个不同的项目合作，是实现人类与自然和谐相处和可持续发展的成功典范。

对柬埔寨领导人的褒奖

在诺罗敦·西哈莫尼国王陛下和诺罗敦·西哈努克太皇陛下的英明带领下，我们赞成柬埔寨人民党取得以下成就：

——把国家从种族灭绝政权中解放出来，在贫困率为100%的情况下，重建国家；

——通过谈判和政治协调，为国家发展和繁荣铺平了道路；

——捍卫宪法，洪森阁下的双赢政策，保证了和平，实现了政治稳定和全国统一，促使了国家经济近几年以两位数增长；

——通过柬埔寨法院特别法庭审判红色高棉最后的领袖；

——派遣柬埔寨部队到苏丹、乍得、中非和黎巴嫩等国家执行联合国维和任务。

在不到20年的时间里，柬埔寨通过一系列政策展示了一种和平与和谐的模式。一种慷慨宽容的方法是建立在包容、慷慨以及确保所有相关人员没有失败者都是成功者的进程上。这种模式与亚洲地区其他冲突相关。以此为背景，亚洲政党国际会议支持，并参加亚太中间党派民主国际，以努力通过在冲突地区加强政党间的对话促进亚洲和平与稳定，例如：尼泊尔、韩国、巴基斯坦、印度和阿富汗。

亚洲政党国际会议根据尼泊尔三个主要党派的邀请，将派代表团去加德满都，寻找三方都关心的并通过对话打破僵局的道路。

亚洲政党国际会议也欢迎阿富汗政府设立国家和平委员会，目的是与暴动者开始谈判，以政治途径平息冲突。

依据亚洲政党国际会议主席何塞·德贝内西亚、秘书长郑义溶给巴基斯坦总统和印度执政党索尼娅·甘地夫人的信函，印度代表和巴基斯坦执政党和反对党于2010年12月3日在金边会谈，并同意在两国主要党派间采取对话机制。

关于近期在朝鲜半岛上的挑衅和军事行动，亚洲政党国际会议呼吁所有相关党派刻不容缓地通过对话和谈判化解当前形势。亚洲政党国际会议强烈敦促国际团体要保证不使用武力，要与联合国秘书长在这一事件上发表的讲话保持一致。

亚洲政党国际会议注意到，"柬埔寨模式"的和平与稳定已呈现出一种解决亚洲其他地区冲突的模式。

致谢

最后，我们必须对柬埔寨人民党和奉辛比克党主办的第六届亚洲政党国际会议表示感谢。此外，我们感谢柬埔寨皇家政府和柬埔寨人民的热情好客。我们感谢德国联邦共和国汉斯基金会和孔拉德基金会及韩国基金会的鼓励和支持。

我们也要对作为观察员参加大会的阿根廷、哥伦比亚、德国、墨西哥、塞舌尔、苏丹、瑞典、坦桑尼亚、乌干达、美国政党的代表，国际生态安全合作组织代表团表示感谢。亚洲政党国际会议向创造性的领导层和充满活力的并作为亚洲政党国际会议总设计师的何塞•德贝内西亚阁下致敬。

附件六：

世界生态安全大会（吴哥）议定书

柬埔寨·金边　　2010年12月3日

本《议定书》缔约方，

我们出席第六届亚洲政党国际会议与首届世界生态安全大会的亚洲国家、南美国家、非洲国家的政党、议会、政府代表团共同关注到，近几年全球性重大生态灾害接踵而来，如地震、海啸、飓风、火山喷发；洪水、干旱、泥石流、森林大火、土地荒漠化、沙尘暴、水资源污染、环境污染、工业污染、湿地锐减、物种灭绝；原油泄漏导致海洋生态系统破坏；再如艾滋病、非典、禽流感、疯牛病、口蹄疫、霍乱等流行性传染疾病，这些由全球气候变化及人类经济活动引发的生态灾害已对人类生存与国家安全构成严重威胁。

《议定书》缔约方意识到，吴哥文明与巴比伦文明、古埃及文明、古印度文明和黄河文明，同是世界文明的重要组成部分。近年来，各国政党、政府在倡导生态文明，应对气候变化，维护生态安全，保护生态环境，实现联合国千年发展目标等领域作出重要贡献。

《议定书》缔约方还意识到，维护生态安全，消除贫困，应对突发性生态灾害，实现经济、社会、生态的协调发展，需要世界各国政党、议会、政府和社会各界的紧密合作，必须采取协调一致的行动。

《议定书》缔约方呼吁，

要加强生态安全理论研究，缔约方应将生态安全纳入国家教育体系，使地球公民认识到生态安全（即生存安全）是地球生命系统（包括大气、土壤、森林、植被、海洋、水等）赖以生存和发展的环境不受破坏与威胁的动态过程。

要坚持生态与经济协调发展。经济发展模式必须与生态承载能力相适应。要把资源永续利用和生态系统良性循环作为重要发展目标，各国制定法律和相关政策要从实际出发，既要考虑生态安全的迫切要求，又要考虑现实可行的经济和社会影响，更要尊重各国国情和自身发展所选择的可持续发展模式。努力促进生态全球化和经济全球化在世界范围内的平衡发展。

要合理分担生态安全责任。当前，世界生态安全状况，特别是发展中国家和欠发达国家的生态环境仍在继续恶化。维护生态安全，实现联合国千年发展目标需要各国共同努力。同时，应当落实"共同但有区别的责任"这一原则。发达国家应积极主动承担生态安全与消除贫困的责任，在生态安全与消除贫困方面发挥更大作用，发展中国家和欠发达国家应将生态建设与经济发展相结合，建立生态安全预警机制，共同应对突发性的生态灾害。

要消除生态技术交流障碍。可持续发展有赖于技术进步，信息、生物、能源等领域的发展，对推动资源有效利用和生态安全具有重要意义。对发展中国家和欠发达国家而言，技术和管理经验的引进与资金的引进同等重要。各国政府和国际组织应当在促进技术合作方面发挥重要作用，既要保护知识产权，又要建立合理的转让机制，促进生态安全体系和扶贫技术的国际交流与合作。

要营造良好的国际合作氛围。建立良好的国际经济和贸易新秩序，需要各国的共同努力。发达国家应进一步开放市场，降低甚至取消因生态安全标准过高而形成的贸易壁垒，促进生态安全与国际贸易的同步发展。发展中国家与欠发达国家应不断加强扶贫力度，提高可持续发展能力，积极参与国际合作与竞争。

《议定书》缔约方还呼吁，

建立国家生态安全领导与协调机构。在当前人口大规模流动和城市化快速发展的情况下，生态灾害随时可能发生。因此，亚洲政党国际会议同意与国际生态安全合作组织及缔约国家共同创建"国际救援机构"，并成立"世界拯救地球家园基金"，通过各种有效途径提高政府、社会、民众的防灾意识和紧急救援救助能力。

《议定书》缔约方同意每两年举办一届世界生态安全大会；设立"维护生态安全特别贡献奖"；同意每两年举办一届世界生态安全博览会；《议定书》缔约方还同意在香港成立"国际生态认证管理机构"，依据国际生态安全管理体系，实施生态安全城市、生态安全园区、生态安全社区、生态安全技术、生态安全产品的认证和管理工作。

《议定书》缔约方认为，防止和减轻自然灾害的一个重要目标是减轻发展中国家和欠发达国家的自然灾害。《议定书》缔约方应广泛开展国际合作与交流，在"国际减轻自然灾害日"举办论坛及宣传活动，将个别国家遭遇到的自然灾害和突发性生态灾害所获取的防灾技术推广到世界各国，如飓风、海啸、洪灾、雪灾、火山喷发、地震、山崩、滑坡、泥石流、地陷等自然灾害的预警、减灾、防灾的成功经验和失败教训等。

　　《议定书》缔约方认为，跨国边境地区的生态系统（森林生态系统、湿地生态系统、海洋生态系统、国际江河流域）的开发，要严格遵循《联合国防治荒漠化公约》《联合国生物多样性公约》《国际重要湿地公约》《国际河流水资源利用赫尔辛基规则》等国际公约规定，做到合理利用、科学开发。各国立法机构要实施生态安全与环境立法，对于破坏生态环境，危害生态安全，导致人员重大伤亡、财产重大损失的机构和责任人要追究经济及法律责任。

　　《议定书》缔约方还认为，本《议定书》或本《议定书》附属文书的解释、适用或发生争端时，要通过谈判或和平对话的方式谋求争端的解决。

　　国际生态安全合作组织是《世界生态安全大会（吴哥）议定书》的执行机构。本《议定书》由第六届亚洲政党国际会议全体会议与首届世界生态安全大会审议通过。

附件七：

蒋明君当选亚太中间党派民主国际
气候变化委员会首席执行官

12月1日，亚太中间党派民主国际（Centrist Asia Pacific Democrats International, CAPDI）在柬埔寨发布《金边和平协议》，呼吁各国要确保本区域和平、安全、稳定、繁荣，各国在处理国与国事务时，采取互不侵犯和干预态度，通过和平手段解决领土纠纷，并遵守国际条约和国际法。

该组织是12月1日上午在柬埔寨总理府会议厅召开的首届执行理事会议时，通过的上述和平协议。大会还选举国际生态安全合作组织总干事蒋明君博士出任CAPDI气候变化委员会首席执行官。

签署协议的政党代表包括尤索夫·卡拉（前印尼戈尔卡党主席）、何塞·德贝内西亚（菲律宾众议院前议长）、穆萨·侯赛因·赛义德（巴基斯坦穆斯林联盟总秘书）、康斯坦丁·格萨乔夫（俄罗斯统一党国际事务委员会主席）、索安（柬埔寨副总理兼人民党常委）、盖博·拉斯米（柬埔寨奉辛比克党主席）、蒋明君（国际生态安全合作组织总干事）、蔡细历医生（马来西亚华人总会会长）、郑义溶（韩国民主党前国际关系委员会主席）等。

"我们提倡裁军和限制大规模杀伤性武器的不扩散。我们也反对一切形式的极端主义、分离主义和恐怖主义。"

该组织承诺维护海事安全，支持《南海各方行动宣言》，并呼吁各方真诚遵守。

此外，《金边和平宣言》将针对联合国千年发展目标，成立亚洲扶贫基金和亚洲小额贷款基金，对最贫穷国家给予债务注销或减免，对环境保护和新经济意识等课题发表看法。

通过全方位中间党合作与对话（如亚洲国际政党会议（ICAPP）、亚洲议会大会（APA）及其他区域政党），以扩大亚洲政党的代表性。

洪森：创造更美好的亚洲

柬埔寨总理洪森出席《金边和平协议》签约仪式时表示，CAPDI的前身为中间派民主国际-亚太区（CDI-AP）。在昨天召开的执行理事会会议上，已通过易名和新章程，以更进一步深化和扩大该组织对不断演变的政治环境进行接触。

"会议通过的新章程，将鼓励更多非政府组织、智库、学者、企业家、媒体、妇女和青年团体的参与，以便促进政治、经济、社会领域的影响。"

"我相信，不同意识形态的政党和国家和平共存，将创造一个更美好的亚洲。在高度全球化的时代，提倡不同政党的互信是必要的。"

他也感谢亚太中间党派民主国际委任他为荣誉主席，并承诺将给予一切力所能及的协助和支持。

通过政党对话寻求和平解决途径

菲律宾众议院前议长何塞·德贝内西亚表示，亚太中间党派民主国际（CAPDI）将关注亚洲各种政治对立和冲突事件，并尝试通过政党对话，寻求和平解决途径。

他接受采访时表示，目前巴基斯坦穆斯林联盟总秘书穆萨·侯赛因·赛义德被委任为该组织的秘书长，他将关注阿富汗冲突问题。

亚洲政党国际会议（ICAPP）创始主席何塞·德贝内西亚指出，CAPDI将与ICAPP合作，为尼泊尔现在的政治僵局和朝韩冲突进行斡旋。

此外，马来西亚马华总会会长蔡细历医生表示，CAPDI是以政党为理事单位的国际组织，因此它可以跨越国界，让不同意识形态的政党进行对话和交流。

"我们坚决反对任何形式的种族极端主义，这在确保拥有多元种族的亚洲国家保持和平与稳定，是非常重要的。"

附件八：

国际生态安全合作组织成为亚洲
政党国际会议观察员

在刚落下帷幕的第六届亚洲政党国际会议上，国际生态安全合作组织被确认为亚洲政党国际会议观察员地位。

近年来，国际生态安全合作组织致力于与各国政党、议会、政府合作，在促进和平发展、维护生态安全、实现联合国千年发展目标方面作出了重要贡献。2009年国际生态安全合作组织参加了在哈萨克斯坦举办的第五届亚洲政党国际会议，并参与起草了《阿斯塔纳宣言》，在气候变化、生态安全、环境保护方面为亚洲政党提供战略咨询，今年国际生态安全合作组织又参加了亚洲政党国际会议在尼泊尔召开的常委会议，并参与亚洲政党国际会议在中国云南召开的扶贫工作会议，目前国际生态安全合作组织已成为亚洲政党国际会议、拉丁美洲和加勒比政党常设会议以及亚太中间党派民主国际（CAPDI）的重要战略伙伴。2010年12月1日在亚太中间党派民主国际会议上，国际生态安全合作组织总干事蒋明君当选为CAPDI理事会成员，并出任CAPDI气候变化委员会首席执行官，柬埔寨王国政府洪森总理，菲律宾众议院前议长、亚洲政党国际会议创始主席何塞•德贝内西亚同时当选为CAPDI荣誉主席和主席。

2010年12月5日，国际生态安全合作组织总干事蒋明君、秘书长郑庆森作为亚洲政党国际会议观察员，与亚洲政党国际会议常委会代表从暹粒乘直升机到柏威夏，在泰国与柬埔寨边境冲突地区视察，并见证了《金边和平协议》的签署，实现了地区稳定与和平发展的良好愿望。

附件九：

蒋明君、郑庆森荣获"为人类
服务特别贡献奖"

　　2010年12月3日，在柬埔寨举办的第六届亚洲政党国际会议与首届世界生态安全大会期间，柬埔寨王国政府总理洪森阁下与亚洲政党国际会议联合主席何塞·德贝内西亚阁下、郑义溶阁下共同签署命令，授予国际生态安全合作组织总干事蒋明君博士、秘书长郑庆森博士"为人类服务特别贡献奖"，以表彰他们在促进和平发展、维护生态安全方面作出的杰出贡献。

　　出席第六届亚洲政党国际会议和首届世界生态安全大会的各国政党、议会、政府代表团共同出席了颁奖仪式。

附件十：

维护生态安全特别贡献奖

根据《第六届亚洲政党国际会议（金边）宣言》和《世界生态安全大会（吴哥）议定书》，由国际生态安全合作组织设立"维护生态安全特别贡献奖"，对维护生态安全，保护生态环境，应对气候变化，实现联合国千年发展目标作出重要贡献的国家政府机构、非政府组织、学术机构以及政府领导人进行表彰奖励。

该奖项每两年评选一次，并在世界生态安全大会期间举行颁奖仪式，首批获奖机构和政府领导人为：

1. 塞舌尔共和国总统詹姆斯·米歇尔阁下；
2. 联合国副秘书长、坦桑尼亚国土住房部部长安娜·蒂贝琼卡博士；
3. 马尔代夫共和国政府；
4. 柬埔寨王国政府；
5. 国际生态生命安全科学院。

首届世界生态安全大会期间，还授予美国堪萨斯市、加拿大列治文市"国际生态安全示范城市"荣誉称号，授予中国东莞市塘厦镇"国际生态安全示范镇"荣誉称号。国际生态安全合作组织总干事蒋明君获国际生态生命安全科学院"学术之星"荣誉奖章，国际生态安全合作组织副总干事、联合国环境规划署俄罗斯委员会项目协调官根纳吉·施拉普诺夫获"全球艾特玛托夫金质奖"。

蒙古共和国前总理、国际生态安全合作组织联合主席门德赛汗·恩赫赛汗阁下，联合国副秘书长、坦桑尼亚国土住房部部长、国际生态安全合作组织联合主席安娜·蒂贝琼卡博士，上海合作组织首任秘书长、国际生态安全合作组织联合主席张德广阁下为上述获奖机构和政府领导人颁发了获奖证书和奖牌。

获奖者简历

詹姆斯·米歇尔：现任塞舌尔共和国总统。米歇尔阁下当政以来，塞舌尔签署了各种与环境相关的国际条约；颁布了宪法保护每一位居民享受并生活在清洁、健康、生态平衡的环境里的权利；向每一位塞舌尔公民传输人与自然和谐共处是大家共同责任的理念，使他们对自然产生天生的热爱，从而在发展的过程中做到自律。这些措施使得塞舌尔与其他地方相比，自然环境被破坏得较少，人与自然保持着亲密、温和、谦逊的关系。

安娜·蒂贝琼卡：坦桑尼亚人，农业经济学博士。于2000年被联合国秘书长安南任命为联合国人居中心执行主任，2002年联合国大会提升了联合国人居中心级别，成立现在的联合国人居署。她在担任联合国副秘书长、人居署主任期间，在政治、财政和运营上大大强化了人居署的功能，扩大了人居署的影响力，使人居署在国际发展中扮演了更重要的角色，推动许多发展中国家走上了消除贫困和可持续发展的道路。2010年9月，她卸任联合国人居署主任。

马尔代夫共和国政府：位于印度洋的马尔代夫被旅行者称做"天堂"，也被认为是地球上最快乐的地方。气候变暖将导致其消失的预言使得马尔代夫对生态环境保持着高度的危机感。为了保护环境，马尔代夫在1989年制定了第一部《国家环境行动方案》，奠定了未来预防环境危机的行动基础，限制了自然资源的过度开采。为呼吁国际社会关注，马尔代夫召开了全球首次水下内阁会议，总统以及11名内阁官员潜至6米深的水下开会，用最直观的行动呼吁各国采取行动应对气候变暖。

柬埔寨王国政府：在柬埔寨政府的领导下，柬埔寨举国积极应对气候变化，改善资源利用状况，落实生物多样性的保护，普及环境保护教育，开展环境评价，开发生态旅游项目。为保护森林资源，柬埔寨制定了《森林法》，明确规定对森林的开发必须有计划地、持久地进行。为保护水资源，近年来，柬埔寨政府采取诸多措施加大对水资源的管理、开发和利用的力度，确保目前和将来均衡用水，保证经济社会的可持续发展。

国际生态生命安全科学院：国际生态生命安全科学院是依据1992年巴西里约热内卢"世界环境发展大会"决议，由独联体国家和国际科学院联合会在俄罗斯圣彼得堡建立的。18年来，国际生态生命安全科学院致力于将世界各国从事生态与生命科学研究的优秀科学家、学者、工程师联合起来，为促进科技进步，进行更为重要、更有前景的科学研究，维护生态与生命安全，保障人类健康作出了杰出的贡献。目前，科学院已与64个国家建立了合作关系，并在纽约联合国总部设立了代表处和常驻代表。

堪萨斯市：位于美国中部的堪萨斯市，一直努力减少对化石燃料的使用，利用当地丰富的风能资源，发展可再生能源经济，提高堪萨斯的能源独立性，保护环境，为广大市民提供一个健康的生存空间。堪萨斯是一个农业大市，为了保护农业生态环境，该市还积极贯彻农业生态保护补贴计划，加大了对农业生态环境的保护力度。扶持农作物生产者实施退耕还林、还草等长期性植被保护措施，最终达到改善水质、控制土壤侵蚀、改善野生动植物生态环境的目的。

列治文市：位于加拿大西海岸、菲沙河入口处的列治文市，一直秉承着保护优先级的原则以及先进的管理理念，应用生态学原理制定城市规划；重视培养人们的"环境意识"和"环境文化"。为了进一步保护生态环境，列治文市大力发展环保产业，加强对资源的综合利用，重视对绿色工业及环保技术的开发工作。在能源方面，列治文市鼓励采用风能、太阳能、潮汐能等清洁能源，鼓励市民选择公共交通或者自行车。正因如此，列治文市才能成为生态安全城市的典范。

塘厦镇政府：地处中国南部珠江三角洲的塘厦镇在工业化和城市化的发展历程中，始终坚持"经济与环境双轮驱动、人与自然和谐相处"的科学发展理念，在全力推动经济发展的同时，高度重视自然生态保护和社会公共事业发展，大力实施"绿色GDP工程、蓝天工程、碧水工程、宜居工程、绿地工程"，从而构筑了塘厦生态绿地环绕、绿色经济循环、和谐文明共存的动态安全生态体系。塘厦镇已荣获"中国国家园林城镇"、"中国绿色名镇"、"中国环境优美乡镇"、"中国卫生镇"等称号。

蒋明君：中国山东省青岛市人，现任国际生态生命安全科学院副院长、联合国青年技术培训组织主席、国际生态安全合作组织总干事。生态安全学博士，1998年提出将"生态安全纳入国家发展"新战略。2006年发起创建国际生态安全合作组织，以维护生态安全，保护生态环境，推进生态文明，实现可持续发展战略为己任。主要研究成果《生态安全：国家生态与发展的基础》、《生态安全学导论》专著，奠定了生态安全理论和学术基础。

根纳吉·施拉普诺夫：现任国际生态安全合作组织副总干事、《国际生态与安全》杂志总编、欧亚作家联盟副主席、俄罗斯作家协会会员。工作之余，他耗时五十多年，终著成诗集《情感生态》。这是一部献给地球上所有相爱之人的杰出诗作，是一曲对爱历久不衰的赞美之歌。

附件十一 * ：

蒋明君专著《生态安全学导论》

蒋明君，出生于山东青岛，现任国际生态生命安全科学院副院长，世界生态安全大会副主席，国际生态安全合作组织总干事，联合国青年技术培训组织主席。

《生态安全学导论》专著凝聚了蒋明君先生14载的心血。在前言部分，他颠覆传统观点，对"生态安全"概念进行了重新定位，指出："生态安全不仅仅是环境保护，而是与政治安全、军事安全、经济安全同等重要，它是人类生存与国家发展的基础。"

全书分为生态安全基础研究和生态安全管理体系上、下两大部分。上部分对生态安全的概念和内涵进行了详细阐述，并重点对国家生态安全进行了分析论述。下部分则从理论走向实践，对国际生态安全管理体系进行了详细介绍，对国内外生态安全城市（镇）、生态安全社区、生态安全基地、生态安全产品的建设管理与评估认证具有很强的指导性与参考性。

＊本文图片为《生态安全学导论》中文版、英文版封面。

附件十二 [*]：

根纳吉·施拉普诺夫诗集《情感生态》

　　根纳吉·施拉普诺夫，俄罗斯人，国际生态安全合作组织副总干事，联合国环境规划署俄罗斯国家促进委员会亚太与中亚地区项目协调官，欧亚作家联盟副主席、俄罗斯作家协会会员，《国际生态与安全》杂志总编，俄罗斯国际生态生命安全科学院院士，全球艾特玛托夫金质奖章获得者。

　　诗集《情感生态》汇集诗人五十多年的创作精华，通过对爱情、人生的感悟，延伸、升华到关注人类，关注地球生态情感。维护地球上的生命是情感生态、心灵生态、希望生态的主题。该诗集的发行所获资金将用于联合国环境保护事业与慈善活动。

　　诗人的第三部诗集《瞬间与永恒》即将出版。

---

　　* 本文图片为《情感生态》封面。

I
Speeches and Congratulatory Letters

Asia's Quest for a Better Tomorrow

Hun Sen, Prime Minister of the Government of
the Kingdom of Cambodia

Excellencies, Ladies and Gentlemen,

It is my great honor and pleasure to be given the opportunity to deliver an opening speech to *this 6th General Assembly of the International Conference of Asian Political Parties (ICAPP)*.

First of all, let me express my warmest welcome to all the distinguished and participants to this important gathering of political parties in Asia and to the 1st World Eco-Safety Assembly. We meet today to continue pursuing the original thrust of ICAPP in promoting exchanges and cooperation between political parties with various ideologies in Asia; enhancing mutual understanding and trust among people and countries in the region; and promoting regional cooperation through the unique role of political parties to create an environment for sustained and shared prosperity in the region.

A fast growing and dynamic Asia has evolved to become the most dynamic region in the 21 century. This has manifested in its phenomenal economic growth and great development potential. It has recently led the global recovery from the worst financial and economic crisis ever. However, as the region is housing several emerging economies and newly industrialized countries, the continent is facing great challenges in coping with rising energy needs to support economic growth and environmental impacts resulted from this fast development. Thus Asian leaders are struggling to strike a balance between economic growth and environmental protection agenda though we are committed to upholding the principle of sustainable development. Therefore, I think, the theme *"Asia's Quest for a Better Tomorrow"*, is very relevant in this context by giving special emphasis on economy, energy and environment issues. I hope that the Assembly and its sideline events will help promote practical cooperation and political commitment to pursuing our shared objectives of sustainable development and finding ways to address the energy needs and environmental impacts.

Excellencies, Ladies and Gentlemen! Organizing this conference on today is no coincidence as today is a historical day for the Cambodian People's Party (CPP). This event is coinciding with the 32nd Anniversary of establishment of the Kampuchea United Front for National Salvation for the liberation of Cambodia from the Khmer Rouge's genocide regime. The CPP has gone through a very challenging period in history during the past three decades —— steering the nation from destruction and devastation toward progress and development.

The Cambodian People's Party is very proud to host this important event of the 6th General Assembly of the ICAPP to show Cambodia to the world that more than 30 years after the collapse of Khmer Rouge Regime, under the CPP leadership, Cambodia is revived, and is moving steadily forward with great confidence and determination; its economic development, institutional capacity, and political freedom are being rapidly strengthened.

Undoubtedly, Cambodian People's Party has been the main driving force in the development of Cambodia. With her full commitment to the battement of the people, started with bare-hands, CPP has been able to make numerous important political, social, and economic achievements in the past 30 years. Owing to the "Win-Win Policy" of Cambodian People's Party that ended the Khmer Rouge regime in 1998, Cambodia has gained full peace, political stability, national unification and territorial integrity. In addition with CPP's right economic policies, Cambodia has achieved rapid economic growth and was one of the fast growing countries in the region with the GOP growth averaged at around 9.3% per year during the past decade, which ensures poverty elevation, and better welfare and standard of living for our people.

Moreover, our progress so far have been achieved owing to the fact that we attach greater importance to regional and international integration and cooperation, starting with the normalization of relations with key international organizations after a very long period of isolation to becoming member of most regional and international forums. Cambodia has now played a more and more important role in the international stage. We have embarked on the training of peacekeeping forces, making humble contribution to regional and global peace. In 1999, Cambodia became a member of ASEAN; and in 2004, Cambodia was the first Least Developed Country to join WTO.

Excellencies, Ladies and Gentlemen! Our gathering today is not strange to the outside world. It is one of a series of high-level meetings held in the region such as the recent ASEAN+3 and G20 summits, which were particularly aimed to dealing with the global financial crisis and the on-going recovery. It should be noted that the recent G20 summit, held in Seoul of the Republic of Korea, demonstrates the increased representation of new political and economic actors in international policy coordination and global governance.

This conference on "Asia's Quest for Better Tomorrow" is very relevant and timely, organized right in the aftermath of the global crisis; where we can see the emergence of a new dynamic regional and global economic architecture – shifting of the center of global economic gravity toward Asia. This trend highlights important structural changes — more opportunities for growth generation through increasing investments, intra-regional trade and demands in the region, but also energy and environmental challenges.

No doubt, it is imperative to exert collective efforts at regional, inter-regional and global level in order to ensure global peace, political stability and balanced economic and social development. However, Asia is emerging as a region that can lead the world economy out of the global crisis and this is no ordinary responsibility. We are recovering more quickly than other regions through domestic demand stimulation and export promotion; growth for Asia is expected to reach 8.2% in 2010, underpinned by a rapid turnaround in exports, healthy private demand, and the lingering effects of expansionary fiscal and monetary policy measures.

In this context, Asia must grab the opportunity and timely uproot the regional and global common challenges. The strengthening of Asia's role must start with the reassessment of sustainable economic growth, energy security, and habitable environment, which, in my opinion, are main factors for sustainable and inclusive development.

Sustainable Economic Growth

Let me underscore the facts that Asia's future economic prospects will depend on the region's success in allowing sources to play a more important role in promoting growth. Joint efforts by Asian countries are needed to rebalance economic imbalances among states within the region and globally.

For Asian region to be less dependent and to expand economic growth base, it needs to diversify the drivers of recovery from export led growth, enabling a larger role for the service sector and private consumption. This requires us to invigorate domestic demand, boosting intra-regional trade through structural reforms aimed at return for investments in domestically oriented sectors and removing impediments and bottlenecks through enhancing business climate.

Narrowing development gap among countries in the region is initially necessary to materialize regional competitiveness and resilience. Strong economic performance in Asia is predominantly led by the economic growth of India and China and robust recovery by South Korea; the region as a whole still has shown broad-based weaknesses. Thus, we need to ensure inclusive growth for Asian countries by enhancing Asia's economic parity and increasing share of benefit – streamline integration, free trade and connectivity. Major priorities for developing countries, such as Cambodia, are to invest more in physical infrastructure and human capital. However, such investments require lots of financial resources. Thus, it is imperative to improve the liquidity management, and financial resources pooling to allow Asian economies to have more control on the ownership of the processes of national and regional development.

Indeed, ASEAN members have shown strong commitment to dosing the development gaps among countries in the region. It has constantly strengthened deeper regional economic integration, and has maintained open regionalism; for example Chiang Mia Multi-lateralization, the ASEAN Connectivity Master Plan and ASEAN Comprehensive Investment Agreement (ACIA) are aimed to enhance Intraregional investment, attract more foreign investment, provide capital sustainability, increase resiliency and link key transport networks. In addition, ASEAN+3 Free Trade Agreement and the ASEAN+6 have already provided formalities for trade facilitation. Further cooperation among Asian nations is economically logical and will help to make this century of Asia. Thus, I think it is important to form is strong economic linkage including bilateral and regional arrangements, initiatives for greater regional integration as well as multilateral trade liberalization among Asian countries. Such cooperation not only strengthens economic development, but also provides a platform for us to tackle other issues such as energy security and global warming.

Environmental Sustainability

Climate change has also posed serious challenges to sustainable development at the global level. There is no doubt that global warming will continue to increase in terms of magnitude and speed. We need to acknowledge that the world is warming due primarily to a human-induced phenomenon. Study concludes that climate change takes place because of the increasing concentrations of greenhouse gases resulting from human activities such as fossil fuel burning and deforestation. Thus, it is an urgent need to find a solution to address the challenges caused by climate change and arrest the emission of greenhouse gases. "Green Economy" must be at the top of the priority list of our development agenda. And we must ensure that climate change is incorporated into all aspects of development policy to promote sustainable management of natural resources. At the same time, there is also a need to increase preservations of environment. The loss of biodiversity is a significant risk factor in economic development and a threat to long-term economic sustainability. Biodiversity is not a luxury, but rather a strategic wealth in securing global economic growth and development.

Developing as well as developed countries will have to seek out a point of common understanding, and start working together to find a resolution to the global warming issue. Constructive consultations, cooperation and increased understanding among countries in the world would definitely help combat the global warming issue. Cambodia fully supports the efforts to address climate change based on the key principles of the united nation framework of convention on climate change, namely "common but differentiated responsibilities and respective capabilities", "specific needs and special circumstances of developing country parties, especially those that are particularly vulnerable to the adverse effects of climate change" and the "precautionary principle". Therefore, Cambodia would like to emphasize on the important linkage between the efforts made by developing countries to address challenges of climate change with the financial support and technology transfer committed by advanced states.

Sustainable development in the Mekong region countries, for example, focuses on an appropriate balance between socio-economic development and environmental protection. More environmental preservation initiatives, projects and studies like "A Decade toward Green Mekong" initiative and the "Hatoyama Initiative" have been introduced to tackle the common environmental problems such as deforestation, forest and land degradation, and water and air pollution caused by rapid economic development and population growth. However, Mekong region countries need to have financial assistance, technology transfer and equipment to ensure adequate capacity for clean energy development, biodiversity improvement, forestry and fisheries management, and pollution control management system.

Energy Security

We all are aware that future economic development will increase more energy demand as well. The world will remain heavily dependent on fossil fuels such as oil, gas and coal for the rest of the century to support development, despite effort of governments to move toward renewable teleology. Studies predict world's energy demand will be doubled by 2050; Asia has grown increasingly more dependent on commercial energy — oil, gas, and coal — for rapid economic development. For instance, China is going to be the world's largest primary energy user in about five years, and India is just right behind China. The price of oil and gas is expected to increase markedly in the near future. Thus, we must do more to use and develop renewable energy sources, as energy security is crucially important in sustaining economic growth and future development.

In addition, I think, cooperation to ensure energy security amongst countries in the region as well as in the globe is necessary; and it can be another beneficial collective effort for Asia, as many of its members are amongst the largest consumers and importers of energy in the world. With higher investment rates in the regions, there is a strong potential for East Asia to move rapidly to the "Green Technology" frontier.

We must also find a fine balance between, on one hand, the development needs of developing countries, especially the least developed countries and, on the other hand, the urgency to address climate change and energy security. Indeed, improved technology efficiency and better use of alternative and renewable energy are also key response to address energy security and climate change. Therefore, we must encourage increasing investment in clean energy development mechanisms.

Greater energy efficiency is vital. So are cleaner energy technologies, including advanced fossil fuel, and renewable energy technologies, which can create jobs, boost industrial development, reduce air pollution and help mitigate greenhouse gas emissions. Increasing energy efficiency is not only good for energy security but is necessary for environmental sustainability. In addition, nuclear option should be retained exactly because it is an important carbon free source of power. Taking nuclear off the table as an alternative option would prevent the world community from achieving long-term goal of controlling carbon-dioxide emissions.

Regionalism — Mekong Region Countries

The recent trends have witnessed the rising importance being given to regionalism. This reflects the mushrooming of FTAs across the globe and the heightened frustration about the slow progress in Doha round negotiations. Thus I would like to share the recent development in the Mekong region which has so far played important role in the Asia's regional cooperation architecture.

The development of Mekong region is necessarily important for ASEAN and East Asian community, as it will reduce the development disparities within ASEAN member countries, and become a pre-requisite of economic integration and securing peace and stability in the Mekong region, ASEAN and East Asia to build a sustainable and harmonized community.

Certainly, the development of infrastructure in the East-West Economic Corridor and Southern Economic Corridor projects aiming at connecting key transportation networks and cross border trade among the Mekong region countries will contribute to economic growth, reducing development gap and promoting broader integration of the region as a whole. So far these development works have not been fully implemented, particularly infrastructure such as bridges and roads of Southern Economic Corridor projects; thus, I would like to appeal to developed countries and development partners to pay more attention to materialize this vision. Financial assistance in a timely fashion is needed, as these projects require huge investment.

Furthermore, living in the era of globalization and fully acknowledging the importance of regional integration, all Mekong countries have collective responsibility for the environmental protection of the Mekong River as a whole. In this regard, to ensure regional harmony and prosperity, all Mekong countries should respect each other mutual interests, by taking into consideration the impacts of projects implemented in one's area can have upon other countries or by making sure that projects put into operation are environmentally friendly for other Mekong countries.

Women Participation and Youth Leadership

Taking this opportunity, I would like also to stress on the importance of women's participation and youth leadership in development. Long-term sustainability and the future of Asia lie on the leadership of next generation, especially those young leaders in Asian countries. In this regard, Asian nations must take a very crucial role in building successors through strengthening and expanding education, employment, and training programs and providing encouragement for wider participation. In addition, social, economic and political development cannot progress smoothly without the important inputs and involvement from women. Thus, there is also a need for active work with women since the process of Asian renovation cannot be implemented without extensive involvement of women.

Parallel to this, we all should strengthen cultural cooperation and exchange. Learning and improving knowledge on culture, custom and dignity of each nation will lead to mutual respect of sovereignty, and contribute to the protection of cultural heritages. Each culture is interestingly unique, and cultural heritage is fundamentally important to national identity. Thus, I am in the view that culture is powerful means to represent who we are, and express our customs and traditions. Therefore, we should embrace cultural diversity and recognize the values and importance of each culture.

Furthermore, to further encourage and strengthen the understanding, and knowledge and information exchange of diverse cultures, it is logically appropriate to establish a "Cultural Association". This association not only helps disseminate information and educate people about interesting aspects of different cultures, but also can help recognize the work of people that contribute to cultural preservation and appreciation.

Before ending, I would like to take this opportunity to profoundly thank the organizers, including the International Conference of Asian Political Parties Secretariat, and relevant ministries and institutions that have worked hard to make this event happen here today.

On behalf of the Cambodian People's Party and Government of the Kingdom of Cambodia, once again, I would like to cordially welcome Your Excellencies, Ladies and Gentlemen to the 6th General Assembly of International Conference of Asian Political Parties in the Kingdom of Cambodia. I hope that this conference will help us to produce the best development path for Asia to seek out a better tomorrow. I would like to wish Your Excellencies, Ladies, and Gentlemen with the four gems of

Buddhist blessing: Longevity, Nobility, Health and Strength.

Finally, may I now announce the opening of the 6th General Assembly of the International Conference of Asian Political Parties.

Strengthen Dialogue between Political Parties and Promote Peace and Stability in Asia

Jose de Venecia, Jr., Founding Chairman of the

ICAPP Standing Committee

H.E. Prime Minister Hun Sen, Father of the Modern Cambodia, and Chairman of the 6th General Assembly of ICAPP,

Prime Minister of Malaysia,

Prime Minister of Nepal,

Former President Megawati Sukarnoputri of Indonesia,

Former President Ramos of Philippines, co-founder of ICAPP,

Vice President Jusuf Kalla of Indonesia, co-founder of ICAPP,

Former Prime Minister of Mongolia,

Minister Li Yuanchao of the Communist Party of China,

Co-chairman Chung,

Excellencies, Ladies and gentleman,

I notice that some of our delegates are with their wives, like myself. I remember the meeting of the political parties in Europe last year, and I asked the Swedish Chairman of the political parties, I said, "Why do you always bring your wife with you during all the conference?" And he said to me, "Mr. de Venecia, I always bring my wife because I always bring the government with me." And so I said in reply, "Yes, you bring the government with you. But in my case, I bring my wife Gina who is the ruling party with me. "

Excellencies, next I'd like to express my appreciation to Hun Sen, conference Chairman, Vice Chairman Sok An, and the political leaders of Cambodia, the King father, and the young king for their outstanding leadership, and of course the

Funcinpec Party for their outstanding leadership to have brought into being this assembly.

As you know, Prime Minister Hun Sen, did his outmost efforts to gather the political leadership in Cambodia to attend to the crisis that happened here and I thank you Mr. Prime Minister for attending to the families of those who finish and of the people who were injured in Phnom Penh last week.

Not only our Prime Minister Hun Sen, and the political leadership in Cambodia found Cambodia's strength at unifying these people of qualifying through the civil war.

Prime Minister Hun Sen, was telling us just an hour ago, that in the 80s when they emerged from the jungle, there were only seven people in Phnom Penh, and he gathered together and forced back into the Jungle in order to nurse this country into a vibrant democracy

Congratulations! Prime Minister Hun Sen.

Prime Minister Hun Sen, Deputy Prime Minister Sok An, the other leaders, also came up with UN, to bring to justice those responsible for the genocidal crimes of convicted on the Cambodian people resulting in the courageous convictions of the mass killer.

Again, congratulations! Prime Minister Hun Sen.

Congratulations to the Royal family and the King Father, the young King, they have restored Cambodia's stability in a way that no other country did, nothing correct but nothing wrong. They are able to do. They have soothed the country's relations with its closest neighbours Vietnam and Thailand.

Recent years, political stabilities have made Cambodia's economy to expand on both global and regional level. Meanwhile we witness the kingdom's economic potential, and the economic recovery has come to the kingdom so quickly.

The new Cambodia reflects Asia's current spirits of our times, from the center of global prosperity it is building towards Asia-Pacific economy and the leader will soon be Asia.

The 21st century must be an Asian century. And of course, our women here today may they say, "Yes, it is an Asian century, but the 21st century will also be the century of women." I can't in front of the delegates disguise this optimism about Asian future.

Ten years ago we gathered in Manila some 50 parties from 21 countries, short of everything we hoped. Not only as our membership increased dramatically, as for this year, we have 318 political parties, ruling parties, operation parties from 82 countries, from northeast Asia, to southeast Asia, to south Asia, to central Asia, to west Asia, to the Arab world, joining me in the ICCAP. We are now increasing the General Assembly in Manila 2000, Bangkok 2002, Beijing 2004, under the leadership of the Communist Party of China, Seoul 2006, and Astana in Pakistan 2009, and that many standing committees leading in these countries.

Even more important we have started up the networks of mutual understanding and friendship dedicated to cooperation among the Asian parties, those in power and those in operation.

Through ICAPP many of our members are developing a national prosperity and the sense of common work that may eventually transcend ideological and regional difference. Creating of this all would certainly acclaim tremendous support from the ICAPP and must receive from the Asian parties and so I believe that just as strong as the reason for ICAPP's attractions is increased the work is also for our commonly strengthened party systems. It's always strong and stable party systems that can bring us tremendous political lasting age, start the shifting of the network, and replace personalized power relations with more durable political associations.

As you know we also began to seek cooperation with the political parties of Latin America, the Caribbean and Africa. Our standing committee has begun to meet regularly the directors of COPPPAL, Permanent Conference of Latin American and Caribbean Political Parties, participating in our first historic meetings in Buenos Aires, Argentina to address our common problems and particularly to present the united strength of global climate change. The political leaders of Mexico are now attending our conference. Let's give them a big clap.

Together the doors of cooperation are also open in communications with the African parties, initially through the African union and two weeks ago when we met in Kenya to establish a triangular partnership in Asia, in Latin America and Africa. Our goal in linking up our three developing continents is finally achieved through the political parties and soon we will reach out to the brothers and friends in Europe and North America. So that someday it can convey a global political party sound.

On a regional and global level, ICAPP's begun to voice Asia's point of view. As initial people of ICAPP, we are grateful that 60 states eventually sponsored the general assembly. Some of our leaderships are considering Malaysia's proposals for financing, anti-poverty climate change programs. Meanwhile the UN's Secretary

General is speaking up for our proposal for an International Anti-Poverty Fund and an Asian Fund from which among others micro-financing institutions could draw. The UN is now also considering to put on the agenda to make ICAPP as an observer to the UN General Assembly but we need resolutions from all your respected governments to present to the UN this agenda as soon as possible next year as presented to the UN by our Co-Chairman Chung.

As we begin our second decade we are also entering a party to party diplomacy, we have communicated with two of our leading members, the India National Congress and the Pakistan People's Party and we also seek ways of easing aspects of the historic, ethnic and religious issues between the two countries ICAPP meetings as our Standing Committee proposed. Prime minister Hun Sen, yesterday talked with us and supported strongly the creation of the peace commission yesterday which is also recommending to ICAPP for your consideration as we have been approved in ICAPP meeting yesterday.

The Northeast Asia Peninsula will also begin exploring the possibility of bringing together in a similar ICAPP context the political parties of the two Koreas. Let us support the political parties of South Korea and the Korean Worker's Party to help them create the beginnings of party to party diplomacy here in ICAPP, here in Phnom Penh.

Such a model has relevant for others conflicts in Asia. In this context, ICAPP supports and joins the Centrist Asia Pacific Democrats International (CAPDI) Peace Commission in its efforts to promote peace in Asia through dialogue among political parties in other conflict areas such as Nepal and Afghanistan.

The ICAPP will be sending a delegation to Kathmandu on the invitation of the three major political parties of Nepal to seek a way out of the impasse through dialogue among the parties concerned.

The ICAPP also welcomes the establishment of the High Peace Commission by the Government of Afghanistan, which is mandated to start negotiation with the insurgents with a view to seeking a political settlement of the conflict.

Asian and global peace must continue to be an alterative role. As a global political religious and ethnic country, we must protect the inferior class.

We also support the efforts of China, Vietnam and Philippines, and the ASEAN country to respond to our problems in South China Sea, we all thank the political parties from China and Japan, and ASSEAN, as we tackle the problem of South China

Sea, we are also in this spirit of party to party diplomacy.

We must also note the new economic and political world order of globalizing and we are developing a system of ideas to ideas in common to suit the globalization's peace content. I firmly believe the system can be more than the sympathies of the public and private partnerships Asia developed that enables private enterprise under development stage to work together. Prime Minister Hun Sen, spoke today of creating, enlarging and deepening the century's political parties, so that we can reach out to the forces of the left and the forces of the right, so that we can support the middle way and perhaps the consensus of compromise among the completing ideologies and parties of the world.

On closing this speech to recall the General Assembly in Manila in September of the year 2000, I noted then that we have come together to write our own chapter in Asia's political history, and it is our wish to build peace, to build freedom, to build prosperity, among all peoples that are gathering all here together.

We saw onto center stage has emerged Asia's economic and political integration. Ten years ago, that goal seemed Utopian, but future generations will witness it. We will be like old men and old women, planting trees for our grandchildren to sit under. We will be acting on behalf of the Asian future. Now Asia's integration has become a possible dream. Both to you and to me it is a reason for our rejoicing through ICAPP's first decade and to be prepared for our second decade.

As we continue to promote political and economic integration in Asia, the tide we see is slowly but clearly emerging, first in southeast Asia's ASEAN, in the Mekong River region, in Northeast Asia, the Asian Parliamentary Assembly, SAARC in south Asia, APEC, Russia and China and Central Asian Regional Economic Cooperation (CAREC), the Arab World's Cooperation Conference, Gulf Cooperation Council (GCC), SASEC and the other great organizations. Yes, the tide is slowly, but clearly emerging. As Asian peoples we say, someday, an Asian community; Someday, an Asian union.

Thank you.

Promotion of Dialogue between Political Parties Contributes to Regional Cooperation

Ban Ki-Moon Secretary-General of the United Nations (Video Speech)

HE Prime Minister Hun Sen of the Kingdom of Cambodia,
Distinguished party leaders,
Excellencies,
Ladies and gentlemen,

Let me start by conveying condolences to the government of Kingdom of Cambodia for the tragic loss of life during last month's water festival. My thoughts are with all Cambodians at this time of mourning.

Ladies and gentlemen, I congratulate the International Conference of Asian Political Parties on its tenth anniversary and for its commitment to promoting dialogue and cooperation in Asia. Political Parties may have very different concerns, context and ideologies, but through regular contacts and exchange of ideas, political parties are also well placed to promote mutual understanding, for more democracy and strengthening regional cooperation. Moreover, they have a responsibility to do this and should be allowed to do this, whether or not they are in power.

These are testing times. The world faces challenges that no nation can address on its own. Climate change, ecologieal degradation, extreme poverty, the continuing impact of economic crisis, nuclear threat, nature and manmade disasters, army violence and violations of human rights. People everywhere live in growing anxiety and fear. However, ICAPP is a support, for the efforts of United Nations for addressing these issues including our work to achieve Millennium Development

Goals.

Thanks as well for promoting women's participation in parties and society, and for helping young people to play larger roles in political decision making. Our work should try to depend on how we work together and for United Nations that can be involved.

Thank you again for your partnership and support and please accept my best wishes for a successful conference.

Political Parties Need to Change with the Times

Dato' Sri Mohamad Najib bin Tun Haji. Abdul Razak,

Prime Minister of Malaysia

First of all allow me to express my gratitude to the International Conference of Asian Political Parties (ICAPP) for inviting me to their 6th General Assembly and the 1st World Eco-Safety Assembly. This is indeed a significant event for me and I am delighted to be here and to be able to visit the beautiful City of Phnom Penh and the Kingdom of Cambodia for the second time this year. I would also like to thank His Excellency Prime Minister Hun Sen and the Government of Cambodia for their gracious hospitality.

Let me also join the earlier speaker to express how sad we are over the recent loss of so many innocent civilians due to a stampede on the bridge of Phnom Penh during the Water festival celebration. On behalf of the people and government of Malaysia, I would like to convey my condolences to the government and people of Cambodia over the unfortunate incident.

My heartiest congratulations also, to the Hon. Jose de Venecia, Founding Chairman and Co-Chairman of the ICAPP Standing Committee, for establishing this very successful forum which brings together political parties of Asian countries, with the aim of promoting exchanges and cooperation between them, in the interest of greater regional understanding and cohesion. Today I am especially honored to be here at the 10th Anniversary of ICAPP, the inaugural meeting of which was held a decade ago in September 2000 in Manila. And I am delighted to be here representing Malaysia, not only as Head of Government but also as the President of the leading component party of the ruling coalition in my country. It is heartening to note that

317 Asian political parties from 51 countries and one territory are able to participate in ICAPP this year. Certainly, the wealth of ideas and experiences coming from the widely varying political parties here represented will make for an interesting meeting of minds.

The theme of this General Assembly of ICAPP is "Asia's Quest for a Better Tomorrow" which is timely, and reflective of Asia's ongoing transformation into the fastest growing region in the world. The forces of globalization and the emergence of China and India as economic juggernauts have changed the way the world views Asia, and indeed how Asia views itself. Social attitudes and mindsets have been altered along with shifting world views and political perspectives, exacerbated by the feverish pace with which technology has engulfed the world. Suddenly every man on the street has an opportunity to broadcast their views and assessment on matters which were previously the exclusive realm of governments and authorities. Suddenly the space for social commentary is wide open and is used and even abused, beyond the limits of common decency and the boundaries of the law. I often say that in politics today, it is no longer as the saying goes business as usual, but in fact it is business unusual.

That being the case, no political party worth its salt can afford to ignore these changes — changes that are affecting the very people whom they exist to serve. Whether a political party stands for traditional views or for unfettered modernism; whether they exist to fight for a specific ideology or to struggle for a cause, the changes affecting society are obviously too important to ignore. The political landscape has altered so radically and whether we like it or not, we have to take cognizance so that acts of political party remains relevant.

A cursory glance at not-too-distant history will show us that many political parties taking into the context of a vibrant political democracy which at the height of their dominance seemed unassailable, but had their fortunes reversed leading to eventual demise, simply because they failed to be attuned with the changing times. It has been quite often said, that power is a heady drug. It can cause among others three things: It can cause inertia, amnesia and even induce delusions. A political party long in power often suffers from these ailments: Inertia in the sense that it does not move forward, remaining static rather complacent and resistant to change and has become too comfortable with its own achievements; Amnesia in that it forgets the original purpose for which it was formed, and the struggle or cause it is supposed to embody; And it suffers delusions in thinking that it's political support base is permanent and unchanging, and in doing so falsely believing that it will forever remain in power. And just because a party that was responsible to achieve freedom for the nation does not mean it will forever in power unless the party itself changes with the time.

Many 'Legacy' political parties are plagued with these illnesses which lead to additional complications of internal bickering and power struggles within the party hierarchy, which more often than not irrevocably damage the party. The end result is the loss of faith of the electorates and party grassroots which cause them to shift their support to alternative political forces that better serve their needs. In the end, the once seemingly unassailable political party suddenly finds itself collapsing under the weight of its own internal issues and maladies, and by then, nothing short of a miracle can reverse its downward spiral.

I raise this point today because I feel it is absolutely critical for any political party to understand the need to change. Not change just for the sake of changing, but change to better serve the need of the party's stake holders; Transformation in the name of greater effectiveness and relevance; Re-invention to remain contemporary and dynamic. Consonant with the theme of this conference emphasizing Asia's quest for a better tomorrow, political parties must begin now to understand that a better tomorrow is founded on an improved political environment in the region.

A pre-requisite towards that end, must be the strengthening of the political process. There has to be greater engagement with the public, to understand their needs and wants in order for political parties to best serve their constituents. Ultimately, contemporary trends indicate that there is an overwhelming demand from the masses for inclusion and participation in the political process and in nation building. A political party that understands that these are the new "rules of political engagement" is the party of the future.

In Malaysia, the ruling political coalition understands this need to change the way we engage our stakeholders, primarily the public at large. We have good times we also have some bumps along the way, not the least of which were the electoral setbacks of the General Elections of March 2008. It was certainly a sobering wake-up call, but in two years since then, in particular the last 18 months, the ruling coalition as a whole has undergone significant transformation and the 13 political parties that make up the Coalition have made great strides in changing and winning back the affections of the people.

Today the ruling coalition Barisan Nasional (National Front) – of which I am

Chairman – continues to re-invent itself to suit the changing times. Most recently, we passed amendments that allowed for direct admission into the coalition, so as to broaden our support base and allow the inclusion of any body or organization not necessarily interested in formally joining any of the 13 political parties in the Coalition. Instantly NGOs and non-profit organizations expressed their keen interest to work with us on the path towards greater progress and development for the country

Within my own political party, UMNO, of which I am the president, we passed historic amendments to the party Constitution last year to allow for greater participation of party members on all levels in voting the ruling party's leadership. Of course it exposes us to greater scrutiny in terms of our work and it certainly puts party leaders including myself at substantial political risk, but it is our belief that our political support must be — and should be — conditional on our ability to perform well, and our ability to convince the people's support now, and by this mechanism, we will be judged by our own party members who can see for themselves whether we have earned our place as their leaders.

Aside from these internal adjustments to accommodate the changing times, the most important thing a political party must remember, as I said earlier, is the constant and continuous engagement with the people. It is this singular element which will be foremost in the minds of the public when they decide whether or not a political party is relevant to them. Real issues about their next pay-check where is it coming from, how will they put food on the table and how will they pay for their kids' education? Such issues are extremely important concerns of the Average Joe. They care far less about the jockeying for positions in the political circles than they do about the pothole on the street outside their home that could damage their cars, or about the broken bridge over a nearby river that would cause problems for them to commute to school and work.

I have always maintained that political parties must be attuned with the pulse of the people. They must be ready to meet the expectations of a society far better informed and able to make comparisons with other societies near and far. It is for this reason that when I took office as Prime Minister in April of 2009, my first order of business was to put in place the People First, Performance Now commitment under the overarching Malaysia banner. It became not only our national tag line but it became our national philosophy. We needed to ensure that we deliver, and are seen to deliver what is important to our people. Since then, we have launched many initiatives such as Malaysia clinics to provide free medical assistance and services to rural areas, Malaysia scholarships for excellent students regardless of race or religion and we aggressively launched various schemes and programmed to improve living conditions of the lower-income groups and eliminate poverty around the country.

On the economic front, the National Front government is relentlessly working to transform Malaysia into a high income economy. In January this year we unveiled the Government Transformation Programmed (GTP) and identified six National Key Results Areas (NKRAs) namely to reduce crime, fight corruption, raise living standards of low-income households, improve rural basic infrastructure and enhance urban public transport. All of these were important areas that have immediate effect on the lives of the public people. Soon after, the Economic Transformation Programmed was unveiled to establish a high income, sustainable and inclusive nation, and 12 National Key Economic Areas were identified to be given special focus to further accelerate Malaysia's move forward. The important thing to note here is that at practically every stage of preparing these economic initiatives which will have impact on their daily lives, members of the public were engaged in public labs and forums that we are able to gauge their expectations and deliver on their aspirations.

Needless to say, the final judgment of whether or not we have met the expectations of the people will only be known from the results of the next General Elections, but early indications are that people have responded positively, and have begun to restore their faith in the National Front. As we begin to deliver on electoral promises and on stated commitments, our credibility is strengthened in the eyes of the public. In addition the people of Malaysia see Malaysia's ruling political coalition as a time-tested, genuine partnership of different political parties striving towards a common goal, as opposed to a hastily cobbled tie-up of ideologically irreconcilable entities solely for political expediency.

I would be remiss not to touch on the external political environment when speaking about a better tomorrow for Asia. Asia is a burgeoning economic zone that is a lynchpin of global trade cannot afford any disruptions or destruction caused by extremism or terror, and in that context Asian political parties have an important role to play.

When I spoke at the United Nations General Assembly in September, I called for a global movement of the moderates to re-claim the center and the moral high-ground that had been usurped from us. I called for moderates to marginalize the extremists and terrorists who have held the world hostage with their bigotry and bias. Nowhere is this more important than in Asia, where in some places cells of extremist groups continue to exist to spread their message of hate.

As members of the Asia family, all of us have a vested interest in ensuring that this region remains free and safe from ideologies espousing conflict, destruction, disunity and hostility in the name of their unholy cause. The proponents of extremism will always attempt to draw lines in the sand, dividing one side against the other, creating the spectra of a nemesis when in fact there is none. I have said repeatedly that the real issue is not between Muslims and non Muslims, but rather between extremists and moderates of all faiths be it Islam, Christianity, Judaism, Hinduism or any other faith.

We cannot allow extremists and terrorists to hijack societal discourse and determine the direction of our respective national conversations. Reason and common sense must prevail. The moderates must always be the only dominant voice, whereas cooperation and negotiation must always be the preferred path over that of enmity and confrontation.

It is in this regard that political parties become pivotal. Political parties, by their very nature represent specific ideals, struggles or causes, for which their members would go to great lengths to defend, propagate and uphold. It is unfortunate that at times, for political expediency, political parties or factions within them would take the path of extremism to strengthen their position. In so doing they would stoke hatred and ignite passionate anger in order to gain short-term political dividends. This is when the center-stage is hijacked and non-confrontational voices are drowned out by angry and dangerous rhetoric. At the same time extreme voices from the opposite end of the political divide would appear, to counter the rhetoric of their opponent, and from that point, in a very short time the country will descent into complete and utter anarchy.

Whatever their causes may be, political parties must refrain from taking the extreme path. There are always peaceful means of conflict resolution, and no short-term political gains are worth sacrificing national and indeed regional peace and harmony. Ultimately what is important is the well being of our people and the preservation of our values, our culture and our way of life. I urge political parties across Asia, to join our call for a global movement of the moderates and reject the politics of hate and any form of extremism.

Indeed, political parties need not resort to extremist posturing to remain relevant and popular. On the contrary, in this enlightened age of information and technology, the most popular political parties would be the ones most visibly responding to changing times and adapting to the wants and needs of the people they serve, as I spoke of earlier. Extremists prey on the fearful and the seemingly threatened. If we make it clear that there is no cause for fear and no looming threat, then extremists become irrelevant.

In calling for a global movement of the moderates, I must emphasize the importance of linkages and networks in furthering the ideals of moderation. Apart from bilateral and multi-lateral government ties, informal networks such as the ones established at forum such as this would be essential in conveying the right message to specific audiences. I strongly encourage all the Asian political parties here to build and strengthen their relationship with each other, exchanging knowledge and experiences, subsequently transmitting this common message, the message of hope and better future back for your people. All of you here are leaders in your community; you are therefore uniquely positioned to spread the word that the key message is one of moderation. If we are able to drive this message home, this message to our people and our communities and it becomes a creed by which they live, then extremism will have truly been debilitated, and a true global movement of the moderates will have taken effect, not just at the leadership level, but right down to the grassroots.

I have shared with you some of my thoughts on what I believe should be the way forward for Asian political parties, domestically and in the context of the external political environment. I have also shared with you some of Malaysia's experiences which by no means perfect, but I hope that the examples that I have highlighted can serve as food for thought.

Every time I visit Cambodia, I recall my first visit here as Minister of Defense in the early 90s during the day of Junta. It was the time where the future of Cambodia was so uncertain and we couldn't imagine the Cambodia of today as we see; this is another success story from where Cambodia came 30 years ago. And that transformation is a credit to the leadership of Prime Minister Hun Sen, who has taken Cambodia from the dearth of desperation to a country that is very stable and progressive. Many people have asked me what the difference is? What makes a country successful and others not so? What is the key difference? It all boils down the one word — "leadership". Leadership that is bold, Leadership that is courage,

Leadership that is visionary and effective. That is the key and Hun Sen has this leadership.

If this is the case for every Asian country then the 21st century will be truly an Asian Century.

Share Experience and View in the Network of Political Parties

Madhav Kumar Nepal, Prime Minister of the

Federal Democratic Republic of Nepal

Your Excellency Samdech Akka Moha Sena Padei Techo Hun Sen, Prime Minister of the Kingdom of Cambodia,
Your Excellency Sok An, Chairman of the Organizing Committee and Deputy Prime Minister of the Kingdom of Cambodia,
Co-Chairs of the ICAPP Standing Committee,
Your Excellencies the Heads of Delegations,
Hon. Ministers and other leaders of political parties,
Excellencies and Colleagues,

We are here on the occasion of holding the Sixth General Assembly of ICAPP, a forum that we created at the beginning of the new millennium. But as we prepare to do so, my thoughts go to the grave national tragedy of the stampede that took place here just about 11 days ago causing a heavy loss of precious lives joyously engaged in celebrating one of the most popular local festivals of this nation of great cultural richness. I would like to convey our deepest condolences to the bereaved families and appreciate the Government of Cambodia for handling this grave national tragedy in a calm and cool manner.

It gives me immense pleasure to recall the fact that in the decade-long history of its existence, the International Conference of Asian Political Parties (ICAPP) has been able to attract and build a forum of an increasingly wider network of political parties in the Asia and the Pacific region. These initial ten years have been crucial for the ICAPP to organize and consolidate itself as a political forum of its own kind. The main theme of the Sixth General Assembly — "Asia's Quest for a Better Tomorrow" — truly reflects our collective commitment for peace and development in the context of the growing dynamism with which the Asian region is making a distinct headway in socio-economic terms in recent years. It is equally remarkable that two special workshops — one each on Women Politicians and Youth Political Leaders — are also being held in cognizance of the increasing importance of women and youths in the political life of a country.

As I look back and take a trip down the memory lane, I find it quite gratifying to recall that we began our journey from Manila in September 2000, and gradually moved ahead through Bangkok (2002), Beijing (2004), Seoul (2006), and Kazakhstan (2009), arriving now in Phnom Penh for ICAPP's Sixth General Assembly this year. As we make progress and consolidate our achievements, my thoughts go back to the formative days in the year 2000 when I had my first meeting and exchange of views with His Excellency Mr. Jose Claveria de Venecia, Jr. which helped to crystallize our thoughts and ideas leading to the establishment of the ICAPP. Today, I am proud to recall that historic point of departure from where we began our long journey together.

I am very glad that ICAPP is no longer limited within the territorial bounds of Asia and the Pacific as we have already developed links with the Permanent Conference of Political Parties in Latin America and the Caribbean (COPPPAL) Coordinating Body. They are here to have the second joint session with the ICAPP Standing Committee following the first one held in Buenos Aires in July 2009. ICAPP and COPPPAL together encompass 68 percent of the world's population, 53 percent of the global GDP, and 50 percent of the total landmass. ICAPP is also equally enthusiastic to develop cooperative links with Africa for promoting global peace and prosperity. The growing popularity of ICAPP as a timely and increasingly relevant political forum, I believe, has been the result of the high spirits, dedication and sense of belongingness of its leaders who broadly define and characterize its core values and visions.

When we started our mission, we were animated by the desire to build a political network for sharing views and experiences and learning from each other's successes in good governance, people-centered development, rule of law, transparency, public accountability, corruption control, democratic institution building and reducing of abject poverty situation. Extensive discussions and deliberations have been held under the aegis of ICAPP on these and other relevant issues including Secretariat's efforts to engage with the United Nations during this period.

Political parties being the most widely recognized democratic instruments for popular mobilization and promotion of the wellbeing of the common people, they have very crucial roles to play in enlightening and empowering the people for building a just, equitable and prosperous human society. It is in this context that the importance of the roles and responsibilities of political leaders need to be understood and assessed. ICAPP has attracted political parties of all hues and colours – leftists, rightists and centrists – from South Asia, South East Asia, East Asia, West Asia, the Middle East and the Pacific as well making it a truly broad-based international political forum. Understandably, the clean image and efficiency of the political party leaders becomes a source of great moral authority. As such, state funding of political parties would help to promote transparency, accountability and value-based politics. We have come a long way in the last ten years in focusing our attention and efforts in these areas. I believe, it has been a great learning experience for all of us to be actively associated with this forum. In an ever-changing domestic as well as international context where we have to work amid numerous challenges of both political and non-political nature, I am of the view that ICAPP continues to provide us a very useful platform for constructive engagement and joint cooperative endeavour for the betterment of human life and general human wellbeing across the region and beyond.

ICAPP has a great role to play in the contemporary world of widening disparities and economic marginalization by engaging in the promotion of global peace, justice and prosperity. Even in an atmosphere of growing interdependence among nations, a large majority of the developing and least developed countries are feeling the brunt of exclusion and deprivation under the current globalization process. This needs to be addressed properly. As current Chair of the Group of the Least Developed Countries, Nepal is especially concerned over the continuing plight of the poor people in these countries who have to struggle every day even for the bare necessities of life.

Excellencies and Friends, may I take this opportunity to share with this distinguished Assembly recent developments in my own country. Nepal has undergone a systemic change and is now in the process of institutionalizing that political transformation by writing a new democratic constitution through an elected Constituent Assembly. We have entered into various peace agreements, ended the decade-long conflict and brought the rebel forces into national political mainstream, enacted the Interim Constitution, elected the Constituent Assembly that is most inclusive and broad-based, consigned the 240-year-old monarchy to history, and have chosen a federal system of governance in place of the erstwhile centralized system.

We still have two major tasks in hand: one, concluding the peace process; two, the drafting of a democratic constitution to institutionalize peace, stability and progress. I have spared no efforts in upholding democracy, rule of law and national reconciliation in my country. We have six months to finish writing the new constitution, and just about a month and a half for the UNMIN to wrap up and return. We have to complete the peace process before their departure. Nepalese people along with all the well wishers and international community are eager to see the peace process completed as soon as possible, new constitution promulgated within the fixed time frame and social and economic progress put on a fast track.

While we work hard to conclude the peace process and promulgate the new constitution, we also make concerted attempts to bring about rapid social and economic development in the country in consonance with the rising aspirations of our people. I am fully optimistic that given the goodwill of the international community and active participation of our people, we will soon be able to turn our country into a proud and prosperous nation that enjoys durable peace wedded to an inclusive pluralist democracy, rule of law and federalism.

Mr. Prime Minister, Nepal and Cambodia enjoy cordial and friendly relations that have good prospects for further enhancement in various fields including tourism and cultural and archaeological aspects. I am glad that as the sacred birthplace of Lord Buddha lies in Lumbini in Nepal, it has provided a strong sense of connectedness between the peoples of our two countries. As a peaceful and peace loving country Nepal is continuously engaged in spreading the message of peace across the world while also actively participating in the UN peacekeeping missions. On matters relating to peace process and national reconciliation, Cambodia is probably one of the most pertinent examples of how a country though devastated by worst form of internal conflict can make use of strong political will to regain national cohesion, unity, stability, peace and progress. Nepal seeks to learn valuable lessons from your experience.

Distinguished Friends, I congratulate the Government and people of Cambodia for hosting this important meeting. I also wish to congratulate the new office-bearers of the Bureau of the 6th General Assembly and wish them all success in taking this forum forward and explore and promote new avenues of co-operation between and among political parties, governments and countries of the region. We have so much of experiences with us individually which we can pool together and utilize for our common benefit. ICAPP does have the potential and the promise to serve us well.

I have no doubt that the Phnom Penh meeting will add a new chapter in the way of further promoting better understanding and fruitful co-operation among the countries, political parties and leaders of our region by identifying new areas of co-operation for our collective advantage. As a large chunk of the populations of most of the countries represented in the ICAPP consists of poor and under-nourished, and have been worst affected by vagaries of climate change, Nepal firmly believes that we should make every effort to focus on poverty alleviation as the over-arching goal of ours.

Thank you.

The 21st Century Calls on Our Immediate Action

Fidel V. Ramos, Former President of the Philippines

Y.E. Prime Minister Hun Sen,
Heads of Government and heads of State,

Thank all the Excellencies for participating in this ICAPP conference. Let me welcome all of you and let us wish life, friendship, good health, longevity, opportunity and a better future for all of us in Asia and the whole world.

The 21st century means we must act now and talk less. For everyone present, the ceremony has gone over for over two hours, so I'd love to shorten my speech, that is, by doing what is 21st century calls for, that is, to talk less and to act now. And you can download my complete speech for the occasion on our website of www.phillipines.betterfuture.com. You got it? (*laugh and applause*). The meaning of hospitable has to be conveyed despite of difficulties and calamities.

Talking less and doing more and what does this mean? It means that we must empower our people. That means education, good health and prosperity of our neighbors. It also means sacrificing for common hood especially for our environment, which not belongs to us, but belongs to the future.

And what does this mean for us as good neighbors? It means industry, sharing and giving for our respectively families, for our Asia Pacific super continent and for the entire world as a whole.

Thank you very much.

Strengthen Dialogue between Political Parties, and Promote Peace, Stability and Prosperity in Asia

Megawati Sukarnoputri, Former President of Indonesia

Prime Minister of the Kingdom of Cambodia, H.E. Samdech Hun Sen,
Founding Chairmen and Co-Chairmen of the Standing Committee of ICAPP,
The delegations of ICAPP,
Honorable ladies and gentlemen,

It's really a great honor to be here in Phnom Penh, and to meet the leaders, and especially represent delegates of political parities in Asia. I would like to take this opportunity to congratulate Cambodia People's Party for its finest hospitality and arrangement that allow us to gather here this auspicious occasion. As you know, Cambodia and Indonesia have shared a long history of friendship since the prime of the ancient Kingdom of Angkor. As a friend of this Kingdom, I would like, first of all, to share my people's condolences and respects for the victims and their families of the stampede during the water festival in Phnom Penh last week.

Excellencies, ladies and gentlemen, my party, our party PDIP, the Indonesian Democracy Party of Struggle, is Indonesia's first party since we have started our democratic transition 12 years ago.

The five principal guidelines of our nation consist of belief in God, humanity, unity of Indonesia, democracy and social justice.

To become the living ideology in the spirit of its first declaration based on Soekarno's speech on June 1, 1945 and in reminding of Pancasila, Indonesia is living proof that all religions and democracy can coexist and prosper and work together in improving social justice and reducing poverty. We appreciate the opportunity to participate in the 6th General Assembly of the ICAPP. We have heard that the ICAPP has become a positive force for promoting understanding among political parties in Asian countries. We believe that dialogue among political parties in the ICAPP always promote the improvement of democracy, peace, stability and prosperity in Asia. Through this honorable moment, as a chairwoman of Indonesian Democracy Party of Struggle, we are now looking forward to be the member of ICAPP.

It would be the great challenge just for us to participate in this Asia's combination, and to bear the bridge of understanding and cooperation among the peoples and the political parties in this region.

Thank you!

Deepen Dialogue and Cooperation between Asian Political Parties for a Better Future in Asia

Li Yuanchao, Member of the Political Bureau of the CCCPC,

Member of the Secretariat of the Central Committee and

Minister in charge of the Organization Department of the CCCPC

Distinguished Prime Minister Samdech Hun Sen,
Distinguished Co-Chairmen Jose de Venecia., Jr. and Chung Eui-yong
Ladies and Gentlemen,
Friends,

It's my great pleasure to attend the 6th General Assembly of the ICAPP and its 10th anniversary celebration in the beautiful city of Phnom Penh. On behalf of the Communist Party of China (the CPC), please allow me to express our warm contributions on the convocation of the meeting. I would also like to express my sincere appreciation to the Cambodia People's Party and Funcinpec as sponsor of the event, and the best wishes to all participants.

The ICAPP was founded at the beginning of this century in response to the global trend of peace and development and the need of Asian countries for strengthening cooperation. In the past decade, with its inclusive approach for cooperation, practical approach for development and innovative approach for creating dynamism, the ICAPP has played an important role in developing relations among Asian countries and safeguarding the regional peace and stability. The gathering has now been developed into an important multilateral forum for equal-footed exchanges, frank dialogues, building consensus and strengthening cooperation between Asian political parties. We are delighted and encouraged by the growth and strengthening of the ICAPP.

Ladies and Gentlemen, Friends, since the founding of the People's Republic of China, particularly since the adoption of reform and opening up some 30 years ago, the CPC has been leading the Chinese people of all ethnic groups in the unremitting endeavor of getting rid of poverty and backwardness and catching up with the global trend of modernization. These efforts are paying off with generally accepted achievements. China, however, remains a developing country. The per capita GDP in 2009 was only 3,800 USD, which was somewhere around 100th place in the world ranking. There are still 150 million people in China living below the poverty line by the UN standard. There is still a long road ahead for modernization in China. In October, the 5th plenary session of the 17th CPC Central Committee was convened, during which the proposals for the formulation of the 12th five-year program for national economic and social development were deliberated on and adopted. Guided by the Scientific Outlook on Development, we will focus our efforts on accelerating the transformation of economic development pattern by deepening the reform and opening up, building up the capability for indigenous innovation, working harder for energy conservation, emission reduction and environment protection and increasing the input for social secret and well-being, therefore promoting the economic and social development in a sound and fast way.

Ladies and Gentlemen, Friends, China unswervingly follows the path of peaceful development. Being an Asian country, the development of China is unattainable without Asia, while the development of China also provides opportunities for Asia. China has always actively participated in various regional cooperation mechanisms in Asia and promoted the regional economic integration. It has provided a helping hand for Asian countries to navigate through the Asian financial crisis in 1997 and the global financial crisis starting from 2008. It's been a cultural tradition since antiquities in China to honor commitment to and seek friendship with neighbors. Peaceful development and a harmonious world are what both the CPC and the Chinese people call for and work for. We follow the principle of building friendship and partnership with neighboring countries and the policy of creating an amicable, secure and prosperous neighborhood in our diplomacy. This line will be followed in the future. China will always be a good neighbor, friend and partner to other Asian countries. The Chinese people will share the opportunities of development and handle the challenges side by side with people from all Asian countries in our unremitting effort to gain lasting peace and common prosperity in Asia.

Ladies and Gentlemen, Friends, with the coming of the second decade in the 21st century, new opportunities and challenges are presented to Asian countries. Against this background, to deepen the cooperation in Asia conforms to the fundamental interests and common aspirations of all the countries and constitutes the shared responsibilities and important mission of Asian political parties. History tells us that cooperation benefits all and strife hurts all in Asia. We hope that the ICAPP could adhere to the open and inclusive approach for further building consensus and enhancing political mutual trust, to the approach of common development for promoting a win-win situation and to the approach of practical innovation for expanding the scope of cooperation, thus playing a greater role in building a prosperous, stable and harmonious Asia.

The CPC very much values the strengthening of cooperation between Asian political parties and always actively commits itself to participating in the ICAPP. As the 90th anniversary comes next year for the CPC, we would like to have theme seminars on issues of common interest with other political parties in due course. We are of the conviction that, with the joint efforts made by all political parties in Asia, the ICAPP will usher in greater development and make greater contribution to bringing Asia a better future.

I wish this General Assembly a crowning success.

Thank you.

Congratulatory Letter

Nursultan Nazarbayev, President of the Republic of Kazakhstan

I am delighted to greet all the participants and guests of VI General Assembly of the ICAPP.

Your activities serve the highest objectives of strengthening friendship and cooperation, trust and understanding among Asian countries, which is especially relevant for joint response finding to the current threats and challenges to global and regional security. We are convinced that true security and development in the Asia may be ensured through wider interaction and constructive dialogue.

Since the earliest days of its independence, Kazakhstan has been continuously increasing its efforts to strengthen regional stability and security. By its own initiative Kazakhstan has voluntarily renounced nuclear weapons and established nuclear free zone in the Central Asia, showing the example way to a safer future.

An active work on enhanced cooperation in the military, political, economic, ecological and human dimension is conducted within the frameworks of the Conference on Interaction and Confidence building measures in Asia (CICA), initiated by Kazakhstan.

CICA gains visible features which allow it in the future to transform into a fully fledged Organization of security and Cooperation in Asia. Interests of peace security and common prosperity underlay Kazakhstan's initiative to create the CICA.

This year Kazakhstan holds chairmanship in the Organization for Security and Cooperation in Europe. OSCE Summit is scheduled for the 1-2nd December at Astana, the same days with ICAPP General Assembly. We believe the forum would give a start to a new stage in the relationship between the East and the West and strengthen mutual trust and cooperation in wide area of Eurasia.

21 century is symbolically called the century of Asia, as the role of Asian states

is steadily increasing around the world. Nowadays Asia becomes the driving force of the global economy and post — crisis configuration of global development is shaped here in the region

I am convinced that the various and complex problems of the region will be resolved successfully though our trust and interaction. The activities of such respected association as International Conference of Asian Political Parties are especially important in this regard.

All the forum participants are wished success and productive work!

Sustainable Development Is the Solid Foundation for a Bright Future of Asia

Hoang Binh Quan, Member of the Communist Party of Vietnam (CPV) Central Committee, and Head of CPV's Commission for External Relations

Mr. Chairman,
Distinguished guests,
Ladies and Gentlemen,

It's our great pleasure to be here in Phnom Penh, the beautiful and hospitable capital of the Kingdom of Cambodia, a traditional and friendly neighbour of the Vietnamese people, together with other political parties to participate in the 6th International Conference of Asian Political Parties (6th ICAPP). On this occasion, we would like to extend our warmest congratulations on the great achievements made by Cambodian people in all areas to build a peaceful, independent, democratic and prosperous Cambodia. We are witnessing a stable, developing and increasingly prosperous Cambodia. We are really impressed with careful preparation and hospitality offered by the Cambodian People's Party and Funcinpec Party for the 6th ICAPP. For this reason, we firmly believe in the great success of this conference.

Distinguished guests, as we all see, this conference is held at a historically significant moment when ICAPP has experienced 10 years of development with many hallmarks. Among those, it is the hallmark made by development with vivid figures: if the first ICAPP marked by the participation of 46 political parties from 26 Asian nations, this ICAPP is featuring the participation of 89 political parties from 36 nations, together with 326 political parties from 52 Asian nations being qualified to take part in ICAPP's activities. It is the hallmark in the relationships among political parties within the region, regardless of how large or small they are, or what political tendencies they take, together with parliament-to-parliament and state-to-state relations, that have been facilitating our ever deeper and broader partnership and cooperation. And it is the hallmark for 10 years of increasingly intensive cooperation and expansive influence of ICAPP. 10 years for an achievement and also 10 years for a future of ICAPP.

Distinguished guests, with an aim to sharing viewpoints, at this conference, I would like to express some of our thoughts:

From our own experience, the first thing we wish to address is that, more than ever, we must highlight sustainable development. As this conference them goes – "Asia's Quest for a Better Tomorrow" – Asia needs a future and has all potentials for a bright future. Asia is the largest and the most populated region, with many potentials in natural and human resources. Having repeatedly rapid economic growth, Asia is the most dynamic region in the world. Asia is rising and growing robustly and vividly. These are vital foundations for a bright future of Asia. However, Asia is also a region characterized by many global pressing issues, including poverty, natural disasters, epidemics, food security and climate change… Therefore, for a bright future of Asia, in our opinion, it is essential to share views and to take responsibility for sustainable development. This is the concern of every nation and obviously of the region as a whole. It is of great significance to understand views and solutions for sustainable development of different nations.

Sustainable development should be a harmonious development of three factors:

To develop sustainably in terms of economy, maintaining macro-economic stability, ensuring economic security, followed by a process of shifting economic structure and growth model; maintaining food security, energy security and secured, efficient performance of financial institutions.

Sustainable development means that economic growth should combine properly the implementation of social progress and equity, continuously improving people's living standards. This means that we have to focus poverty reduction and filling the gap in living quality. Given the experience and success of our Party and State as the pioneer in hunger elimination and poverty reduction, which has been recognized by international community, we are willing to share our practical experiences with other political parties and nations in this field.

Lastly, sustainable development should lay special emphasis on environment, economic development must be closely linked with environmental protection and improvement; proactively responding to environmental disaster; applying material and energy-saving technologies. This is clearly the concern of every nation and also responsibility of one nation towards others, as well as for sustainable development of the whole region.

In this sense, sharing views on sustainable development, followed by cooperation is a very important factor. Therefore, we recommend ICAPP to have regular mechanisms of exchange such as workshops and seminars on "sustainable development" in order to raise awareness among Asian political parties and to thoroughly grasp views of sustainable development in investment, trade, exploitation and consumption of energy.

Distinguished guests, Asia is performing dynamic development, however it is also featured with strategic engagement and competition of interests; Asia's future needs a foundation of stability and peace. With such context and requirement, we suggest that Asia needs to have and to be enhanced with exchange mechanisms on regional security, in which the role of Asian political parties should be promoted.

Asia needs future and will have its future, thus all of its potentials need to be awaken for the future. Asia is the most populated continent and we believe you would share our view that "People are the power". Therefore, people-to-people exchange should be emphasized and presented in thoughts and activities of political parties. In this context, we would especially like to mention about youths. In any nation, youths are always a major force and play decisive roles to the future of each nation-state, therefore, youths have indispensable role to the future of Asia. We would propose that political parties need to consider establishing a regional youth exchange mechanism, and based on such results of vital reality, to further consider establishing a regional youth organization. At first, we suggest to have a form of exchange among young politicians to strengthen mutual understanding, sharing and learning. In terms of organizing, this is important foundation for closer cooperation among political parties in the future.

Ladies and gentlemen, we would like to brief you about our country. Nearly 25 years of renewal have brought about a new face to our country with socio-economic stability; people's living standards have improved remarkably; economic growth rate have been maintained relatively high and continuous for a number of years. Achievements that we have gained not only meet the expectation of Vietnamese people from various strata but also conform with our commitments to the UN Millennium Development Goals.

The 11th Party Congress of the Communist Party of Vietnam is to be convened in first half of January 2011. The Congress is a significant political milestone with a focus to review, supplement and develop our Party's Political Program and map out strategic plan for socio-economic development in the next 10 years. Our target is by 2020 Vietnam is expected to become a modernity-oriented industrialized country and by the mid-21st century a modern industrial, prosperous, strong, democratic, equitable and advanced country.

We consistently implement the foreign policy of independence, autonomy, peace, cooperation and development; multilateralize and diversify international relations, be proactive and active in international integration with the spirit: "Vietnam is willing to be a friend, reliable partner and responsible member of the international community".

Once again, we would like to strongly assert the wish of the Communist Party Vietnam to further strengthen relations of cooperation and friendship with other political parties in the region and the world. We would be very happy and grateful to receive messages of congratulation from international friends to the 11th National Congress of our Party.

As a member of ICAPP Standing Committee, the Communist Party of Vietnam is deeply aware of its role and responsibility in promoting ICAPP common goals, and willing to share common opportunities and challenges. We would look forwards to further enhancing understanding, confidence and cooperation among political parties in the region, together to maintain peace, stability and prosperous development in the meeting of ICAPP Standing Committee in Vietnam in the coming time.

In conclusion, may I wish the 6th ICAPP a great success.

Thank you for your attention!

We Are Responsible for Bringing Stability and Peace to Our People

Mohammad Azam Khan Swati, Special Envoy of the President of

Pakistan

Excellencies:

First of all I would like to express my sincere gratitude to all Honorable parliamentarians, political Leaders and Guests for taking time out of their busy schedule and engagements to attend this conference. Congratulate ICAPP and its visionary leadership for attracting competing governments, opposition and independent political parties of this region and beyond for such August forum.

Ladies and gentlemen, I am certain that Asian leadership of today have gathered here under one roof with a theme of Asia's quest for better tomorrow. Realizing that progress in any human task is only possible if we can bring peace to our respective countries, which is indispensable for our progress, prosperity and survival.

Afghanistan which is gateway to Asia is facing a war for thirty long years. This unwarranted bloodshed has created an environment for its citizen to fight back and defend themselves rather to think of peace and Tolerance. Now Pakistan and other neighboring countries are threatened to maintain their peace and security. On the other hand world powers in the name of fight against terrorism are trying to occupy not only the natural resources of Asian countries but also to subjugate their political system, their sovereignty, their financial and monetary system.

Therefore, it is incumbent upon you as a leader to harmonize your policies that can bring stability, peace and socioeconomic wellbeing to your people. It is also your moral duty to mutually resolve basic issue confronting south Asia countries among themselves. To me it is the issue of Kashmir between India and Pakistan, which is pending per united nation resolution for sixty long years.

Given the sweeping changes all around us world has well recognize century for progress, prosperity and innovation, This requires us to adopt new solution to resolve problems, invest in human capital, bring a positive change in our socio economic Structure, invest in health, education and environment, alleviate poverty, invest in higher education research and Technology that can bring equality to all human being where a common men or women get his equal share of opportunities and can in turn enjoy safety, security and viable environment to reach his or her destiny. In the end I hope that deliberations of this conference and the interaction between delegates of the region would signification help in the formulation of appropriate policies, frame work and a dawn of better tomorrow.

Thank you all.

Underline People-Oriented Development Concept

Jean-Paul Adam, Foreign Minister of the Republic of Seychelles

Y.E. Prime Minister Hun Sen,
Your Excellencies Ministers of Cambodian governments,
Your Excellencies participants of ICAPP leaders,
Ladies and gentlemen,

First of all, thank you very much for welcoming Seychelles to this forum. I should tell you about us since we are probably not known to many of you.

Seychelles is a small group of islands in the middle of Indian Ocean, and I would like to tell you that, for Asia, we are your first stop on the way to Africa. We are the island that can connect you to the African continent, and it gives us really a great pleasure to be here today. For hundreds of messages of my president and I also would like to say that you probably also noticed that I'm absolutely young for a foreign minister, but will you please to note that, Your Excellency the prime minister of Cambodia, was also elevated to power, or was in politics in a very young age. And I believe that I'm very inspired by the Prime Minister. Do you think that in Asia we can see that Asia has been able to pioneer the power of youth, and I think that the future for Asia and for Africa is continue to harness the power of youth and turn to positive development for our peoples.

Your Excellency, on behalf of the President of the Republic of Seychelles, His Excellency, James Michael, I would, first of all, like to convey the condolences of the people and government of Seychelles to Kingdom of Cambodia for the unfortunate event occurred during the water festival. We're very encouraged by the responsive movement of Cambodian government, which was very quick to deal with the very difficult circumstances and which has also brought the country together at this moment of difficulty. I would also like, on behalf the President, to convey to you that we really are impressed by the development of ICAPP and we feel that as a smaller country in Africa and also as a member of African Union, Africa also has a lot to learn from Asia, and a lot to learn from its experiences over ICAPP. We have learned a lot of the need of concerted building and how to successfully use it as a tool for democracy and for good governments in this region. We also feel that in ICAPP there is no country of ideology and this is something which I think Africa can also benefit a lot from. I feel something which has been very positive in Asia is the concept of putting people at the center of development, and this is something which is key to development. And I think I would like, on behalf of President Michael, to convey to government of Cambodia, our congratulations of successfully putting people at the center of development.

The future for Asia and also for Africa also depends on that be able to pursuit sustainable development with the ecological safety and environment protection as its core. So, it is opportune at this moment of ICAPP is being held as well as the Eco-Safety Assembly. We note that climate change is one of the greatest senses by mankind. And as a group of small islands developing states, climate change can absolutely affect the way of our development as a nation. Having these two events side by side, the popularization of ideas allows us to make sure that the environment is at the center of the political agenda. So, we also would like to convey our support for the statement made by H.E. Prime Minister Hun Sen on climate change. We believe that we need the common but differentiate responsibility which is the key to find the successful solutions to this problem. So, thank you for your leadership on this matter and for your support. Your Excellencies, ladies and gentlemen, my speech may be too long, thank you for your attention. On behalf of my government, the people and the President of Seychelles, I reiterate our sincere appreciation and our congratulations on this fantastic Assembly.

Thank you very much!

Strive for Building our Home Planet to be Ecologically Safe and Sustainable

Keo Puth Reasmey, Deputy Prime Minister of Cambodia, and Executive Chairman of IESCO

Distinguished Guests,
Ladies and Gentlemen,

Good morning!

Today we are gathered here in Phnom Penh for the 1st World Eco-Safety Assembly. First of all, in the name of the hosting country, and on behalf of the Royal Government of Cambodia, I would like to extend my warm welcome to all the distinguished guests participating in the Assembly. At the same time, I would like to express my sincere appreciation to the Sixth International Conference of Asian Political Parties and International Eco-safety Cooperative Organization for your wisdom, creativity and passion to gather us here today! My appreciation also goes to the staff and volunteers of all sectors of Cambodia who have made efforts for the successful convening of the Assembly, for your dedication, participation, understanding and support, which have made it possible for us here to be able to probe into the development plan of global ecological safety.

20 years ago, the United Nations Secretary-General Boutros Boutros-Ghali pointed out that climate change is a fact, and that we must make changes and sacrifices — we can not live any longer at the expense of the interests of our Earth and future generations. Since then, we have asked ourselves: in response to climate change and global ecological crisis, have we done enough? Clearly the answer is NO! That is the reason for our reunion here today.

As is known to all, safety in its traditional sense is mainly restricted to military level and diplomatic level, however, with the relaxation of international situation, and particularly in the time of peace, people have gradually recognized that it is far from enough to discuss safety only at military level, because in today's world, the issue of safety confronting every country is not just decided by element of military, but also that of economy, energy, resources, environment and population.

Recalling the history of industrialization, we can see that most countries have adopted the strategy of "economy priority" and have generally undergone a development path of "pollution first, treatment later", which is a very painful price people have paid in the process of the development of industrial civilization.

At present, ecological safety has become an important factor in coordinating economic development for all the countries in the world, and energy security and climate change have become the world's most pressing environmental issues. The world countries are gradually incorporating ecological safety and environmental protection into their national development strategies, and have made certain achievements.

Since 1960s, Cambodia and Southeast Asian countries have made considerable progress in economic development, but subsequently confronted with the intensification of environmental pollution and deterioration of ecological environment, such as urbanization, air pollution, water pollution, and reduction of the tropical rain forests, with trends of cross-border pollution. Deterioration of ecological environment has posed a serious threat on the survival and development of Southeast Asian countries.

Cambodia is one of the countries that are most vulnerable to climate change in Southeast Asia. Years of continuous flooding have brought great impact on the sustainability of Cambodia's economy and society. Abnormal rainfall, drought, desertification, floods and other natural disasters not only exert direct impact on the safety of the entire country, but also affect the safety of crops, food, water and so on. Facing the adverse effects on national economy resulted from ecological imbalances and climate change, we have recognized the seriousness of the problem, and we are going to enhance the environmental awareness of citizens, strengthen environmental protection, handle the relationship between environmental protection and economic development, and enhance the international cooperation in the field of eco-safety, all of which are urgent priorities of governments at all levels.

Today, we have recognized that improper exploitation of human environment and large amount of greenhouse gases emitted by industrial countries are the main factors causing the deterioration of ecological environment. Only by protecting ecological environment can we protect our right of survival and development. It needs the efforts of the entire humanity to cope with climate change, maintain ecological safety, protect ecological environment, and tackle unexpected natural disasters and ecological crisis!

Here, I would like to express my sincere hope that this Assembly will provide an opportunity for all the countries in the world to strengthen the cooperation committed to solving global ecological crises, and that the thematic sessions of the Assembly will propose specific programs to promote ecological safety and sustainable development.

Ladies and gentlemen, dear friends! Air, sunlight, water and dew have given birth to a vibrant Earth, which is the basis for the development of mankind. Let us be full of sense of mission, share responsibility and strive for building our Earth Planet to be ecologically safe and sustainable!

Finally, let me finish by wishing the 1st World Eco-Safety Assembly a great success!

Congratulatory Letter [*]

Mohamed Nasheed, President of the Republic of Maldives

ICAPP gathers in Cambodia during a year in which the world has suffered one climate-related disaster after another: floods in Pakistan; fires in Russia; and severe storms in the United States.

2010 is predicted to be the hottest year on record — next year will probably be hotter still. Yet the world has still been unable to come to an agreement on climate change.

We must – and we can – do better.

I believe we need to view climate change not just as a challenge but also as an opportunity. Cutting carbon should not be considered a burden that will destroy jobs and hamper economic growth. Instead, going green should be seen as the greatest economic opportunity since the Industrial Revolution.

This is an opportunity to improve things to grow our economies in more sustainable ways and to create new wealth and employment.

Progress on climate at the Cancun Climate Change talks – and in South Africa in 2011 – should be viewed, not as an impediment to growth, but as a boost. A deal must not be seen as a drag on development, but as a way of doing things better.

The Maldives aims to become carbon neutral by 2020. We make this commitment not because we are 'tree huggers' – in a country that is 99% sea, there are few trees to hug. We have made our carbon neutral pledge because quitting fossil fuels is in our economic and energy security interests.

Electricity generated from fossil fuels in the Maldives is extremely expensive. And our vulnerability to the unpredictable price of foreign oil places our economic development at risk. For the Maldives, we need to adopt renewable energies in order

---

[*]   This congratulatory letter was read out by Ahmed Latheef, Special Envoy of the President of Republic of Maldives.

to develop. I suspect the same is true of many other countries represented in this room.

I believe ICAPP can play a great role in safeguarding the climate and helping Asia lead the renewable energy revolution. I appreciate very much the role ICAPP is playing to bring together political parties in Asia.

Only through unity can we solve our common threats and take advantages of new opportunities. Organizations such as ICAPP are crucial in bringing us together and helping us bridge the transnational challenges of the 21st century.

I wish you every success for your meeting.

Push forward Development of Eco-Safety through Continuous Innovation

Congratulatory Letter of James A. Michel, President of the Republic of Seychelles, and Speech by Jean-Paul Adam, Foreign Minister of the Republic of Seychelles

Your Excellencies, Ladies and gentlemen,

It is a great honour for me to have the opportunity to address you today on behalf of the President of the Republic of Seychelles, James Michel.

President Michel unfortunately could not travel to this prestigious event because he has gone to Cancun to attend the United Nations Climate Change Conference, and he has asked me to convey on his behalf, his sincere appreciation to the Government and People of Cambodia for bringing the concept of Eco-safety to global attention by hosting the First World Eco-Safety Assembly.

May I also add my own feelings of gratitude and appreciation to our hosts for welcoming us with such warmth and hospitality.

I must also take this opportunity to convey on behalf of the President and people of Seychelles, our most sincere condolences to the Government and People of Cambodia following the tragic losses sustained during the stampede that occurred last week during the water festival. We note how the people of Cambodia have come together in this moment of great loss, and we also stand shoulder to shoulder in solidarity with you at this time of great mourning and grief.

Your Excellencies, ladies and gentlemen, President Michel, as a leader of a small island Developing State, has always sought to achieve more understanding and cooperation in the international community with regards to the importance of environmental protection as well as eco-safety.

I will now share with you the message that he would like to address to this summit, on behalf of the people of Seychelles:

Excellencies,

Ladies and gentlemen,

We are currently standing at a cross-road of global governance.

We are faced with the ever increasing threat of climate change, while the climb towards development continues to get steeper. The existing means at our disposal to tackle these challenges are often too slow and too inefficient.

The World Eco-Safety Assembly offers a new commitment to address these issues in a concrete manner. The concept of eco-safety highlights the fact that the world's security is inextricably linked to our ongoing efforts to achieve true sustainable development.

In islands such as Seychelles, the link between security and ecological protection is very obvious. Our tourism and fisheries industries depend on the sustainability of our natural surroundings. Our wealth can only be guaranteed if we protect the source of our wealth- the environment.

This is why Seychelles has continued to increase our protected areas. Earlier this year, I declared our Silhouette Island, a nature reserve. This brings the total amount of protected land in Seychelles to 47% of our land territory — the highest proportion in the world.

As we assess the results of the discussions on Climate Change in Cancun, also we look to the World Eco-Safety Assembly, to play a role in promoting the concept of "common but differentiated responsibility" between developed and developing countries, in relation to resolving the climate challenge.

If we take the example of small island states — they are the ones with the most to lose from climate change while having contributed the least to its causes. Small islands are also often marginalized in the international development architecture. We have limited access to development funding because of our high GDP per capita. Many small islands are also highly indebted and struggle to achieve their development priorities without recourse to expensive commercial credit.

And as we struggle to adapt to climate change and its associated costs — the sustainability of our economies is further weakened.

Eco-safety is very much about finding the right model to preserve the last paradises on earth. Cities can be rebuilt. But no one can replace the miracles that nature shares with us through its oceans, its sandy beaches, its coral reefs, or its myriad of species that inhabit them.

So while the developed world is right to be concerned about troubled economies, the true challenge of sustainability lies in our natural environment. If we cannot save our environment first, eventually there will be no economy to save. This applies as much to the world economy as it does to small islands. The only difference is that islanders see this as part of our daily lives. While a large part of the world's population is divorced from the environment which actually sustains it.

Part of the concept of eco-safety is about our understanding of environment. We need to reconnect with the ecological environment on the global level. Our World Eco-Safety Assembly can play a key role in achieving this.

Part of our efforts also needs to focus on the drive towards making renewable energy and the technology that goes with it accessible to developing countries.

Small islands have an abundance of potential to develop renewable energy such as solar or wind energy, but our access to the latest technology is often limited.

We need to be innovative in our approaches towards these developments. There is also significant scope for Private Public Partnerships (PPP) in this regard and we should not shy away from tapping into commercial possibilities of renewable energy. If harnessed properly, this can provide a new developmental impetus to developing countries.

Excellencies, ladies and gentlemen, The World Eco-Safety Assembly is the ideal forum to create a new momentum for the sustainability of our planet. We need to bring together many ideas which on paper may seem incompatible. We need to grow, while also preserving. We need to meet immediate needs, while also thinking of the future.

We are pleased that this forum has brought together many of the finest minds with innovative ideas to address these incompatibilities.

I would like to particularly thank Professor Jiang Minjun, for working diligently around the clock to ensure that this Assembly became a reality. The drive to bring us together, reflects the desire for us to move beyond the tried and tested- to truly put eco-safety at the centre of global governance.

The first obstacle, the first major issue which is incompatible with sustainability, is for us to continue with the status quo. But I am confident that the World Eco-safety Assembly can change this, and by doing so, we can change the world.

I thank you for your attention, and hope that you will all have the opportunity to also visit Seychelles and discover for yourselves, the beauty of the Indian Ocean.

Excellencies, ladies and gentlemen, it has been my distinct pleasure to address you today on behalf of the President. His message reiterates the commitment of Seychelles to the ideals promoted by this Assembly.

We look forward to the outcomes which we believe can make a real difference.

I thank you for your attention.

Congratulatory Letter *

Yoweri Kaguta Museveni, President of the Republic of Uganda

Your Excellency Prime Minister Hun Sen,
Your Excellencies, Former Heads of State of Government, Secretary-General of ICAPP,
Distinguished leaders and members of ICAPP,
Distinguished guests, distinguished delegates,
Ladies and Gentlemen,

Allow me, Y. E., to convey to you very warm and gratitude greetings from H.E. Yoweri Kaguta Museveni, President of the Republic of Uganda, the government and people of Uganda. Allow me also to convey to you, the government and people of Cambodia, heartfelt condolences and sympathies from the government and people of Uganda upon the tragic lost of lives that occurred at the water festival.

Yesterday, Y. E., I and other delegates had an opportunity to visit the genocide museum at Toul Sleng. The museum that is living testimony of the tragedy that befell your country in the 1970s, the testimony of Idi Amin committed by our government that was supposed to defend and protect its people. Y. E., Uganda and Cambodia share a common history of instability and bloody shed. And interestingly, both occurred about the same time and ended it about the same time in the 70s. Uganda witnessed Idi Amin, between 1971 and 1979, and lost many people under leadership that was supposed to properly defend and protect them. And as Uganda, Cambodia has risen from the ashes, under the current leadership of Prime Minister Hon Sen, and Cambodia People's Party, and the government, to transform Cambodia into an icon of stability and rapid economic development. Allow me, Mr. Prime Minister, to congratulate you profusely and your government upon these spectacular achievements. May I also congratulate ICAPP on the 10th Anniversary and for the

---

*   This congratulatory letter was read out by Madibo Charles Wagidoso, Special Envoy of President of Uganda.

achievements made in strengthening cooperation between political parties in Asia. We welcome efforts to expand this cooperation and strengthen linkages with political parties in Africa, and indeed, in other parts of the world. We also look to the day, in the words of H.E. Jose de Venecia, when he said we look to the time, when we have the conference of global parties. The success of ICAPP is indeed an inspiration for the developing world and shows that, together, we can achieve our common purpose for a better life.

Finally, Y.E., I'd like to thank you for inviting Uganda to attend this conference, and also thank you in particular, Prime Minister Hun Sen, the government and the people of Cambodia, for the warm hospitality accorded to us, and for successfully holding the 6th General Assembly of ICAPP in this beautiful city of Phnom Penh.

Thank you all.

Congratulatory Letter *

Sergei Mironov, Chairman of the Council of Federation of the Federal Assembly of the Russian Federation

Respected Mr. Jiang Mingjun and dear friends,

On behalf of The Council of Federation of the Federal Assembly of Russian Federation, I would like to express my gratitude to all participants in the First World Eco-Safety Assembly.

Your activity, no doubt, is dealing with a critical problem that all humans are facing, which is eco-safety. Natural environment provides a pre-requisite factor for human survival. But we sometimes abusively use natural resources to satisfy our own needs.

Vernadsky, a famous Russian scholar, in his book pointed out that human being is a critical "geological factor" on the evolution of earth surface change. Our mission nowadays is to ensure that the environmental change will not threaten human itself, and the entire nature. We now need to adopt a common measure to strike for biotic environment balance, and to recover the natural environment that is suitable for human living.

I firmly believe that the first Eco-Safety Assembly will make a great contribution toward improving our efficiency and effectiveness on the work of environmental protection, and toward developing clean industrial technologies etc.

Wish all of you productive work, and success in professional and innovation work.

---

* This congratulatory letter was read out by Keo Puth Reasmey, Executive Chairman of IESCO, Deputy Prime Minister of Cambodia.

Ecological Safety Needs Our Common Care

Madhav Kumar Nepal, Prime Minister of Federal Democratic

Republic of Nepal

Honorable Chairman of the Assembly,
Distinguished guests,
ladies and gentlemen:

Good morning! I feel very honored to attend the 6th General Assembly of the ICAPP and the 1st World Eco-Safety Assembly. The theme of this Assembly is "Peaceful Development and Ecological Safety". I believe, like me, that the delegates present here all pay close attention to eco-safety problems and wish to push forward peaceful development and jointly protect our homeland—earth through our own humble efforts.

Seeing from the perspective of international relation, national contacts featuring environment and ecology can be dated back to 1970s. Due to the increasingly prominent environmental problems, a green movement of large scale sprang up around the globe in 1972. United Nations Conference on the Human Environment, which was convened in Stockholm in June of the same year and during which Declaration on the Human Environment was issued, marked the beginning of world "environmental diplomacy". In June, 1992, over 180 heads of state gathered together in Rio de Janeiro, Brazil to convene United Nations Conference on Environment and Development. This conference formulated Agenda 21, signed United Nations Framework Convention on Climate Change and Convention on Biological Diversity, issued Rio Declaration on Environment and Development and Forest Principles. This conference had become a major milestone of "environmental diplomacy". In December, 2009, Conferences of Parties to the United

Nations Framework Conference on Climate Change was convened in Copenhagen, and over 190 heads of state attended the conference. On the occasion of the Assembly, the new Conferences of Parties to the United Nations Framework Conference on Climate Change is held in Cancun, Mexico.

Seeing from the global scale, ecological safety has become the theme of the present era to which the whole world pays close attention. Everyday, through various media, we can see the reports on natural disasters and unexpected ecological disasters in different places of the world such as drought, flood, typhoon, storm surge, freeze injury, hail, tsunami, earthquake, volcanic eruption, landslide, mudslide. Apart from natural variation, human activities are the reason leading to those disasters. Maintaining ecological safety and protecting natural environment need our joint efforts.

Located at the southern foot of the middle section of the Himalayas, Federal Democratic Republic of Nepal is famous as a mountain country with mild climate and beautiful landscape. It is known as "beautiful scenery of Mount Everest, holy land of Buddha". Nepal has always attached great importance to the protection of ecological environment. Form tundra to prairie, tropical rain forest to shrubbery, Nepal has established many wildlife protection zones, national parks and protected areas. Therefore, tourism is one important economic pillar of Nepal. Comparing with other regions in Asia, Nepal has managed to protect many endangered animals and plants. At present, the Ministry of Forestry and Land Resources is carrying out the eco-safety and sustainable development project of Nepal, which aims to protect forest reserves, call attention to regions with biological diversity, promote the protection and development of wetlands, initiate the management plan of community natural disaster, strengthen management of climate change and natural resources, enhance urban air quality, improve hygienic environment, develop ecological energies, and establish sustainable mechanism of resources on the basis of sharing gains and incomes.

In 2008, a delegation of IESCO visited Nepal for a fact-finding trip and then decided to, together with the Ministry of Forestry and Land Resources, establish "Lumbini International Eco-Safety Demonstrative Base", aiming at pushing the construction of international eco-safety natural reserves, and the projects of eco-safety and urban sustainability. We also invite other international organizations and non-governmental organizations visit and inspect Nepal and give support to our efforts to build a country with ecological safety and sustainability.

Thank you!

Push Harmonious Development between Human Being and the Earth by Using "3E Concepts"

Anna Tibaijuka, Former UN Under-Secretary-General and

Minister of Lands, Housing and Human Settlements

Developments of Tanzania

Respected Chairperson of the Assembly,
all the distinguished guests,
ladies and gentlemen,

Good morning!

I feel much honored to come to this beautiful Phnom Penh, capital of Cambodia, to attend the opening ceremony of the First World Eco-Safety Assembly. Thanks to the arrangement of organizing committee of the Assembly, I have the opportunity to discuss with all your distinguished guests who are present a common topic and an extremely urgent matter, namely, the problem of eco-safety and sustainable development.

During my term serving as the UN Under-Secretary-General and Executive Director of UN-HABITAT, my team and I proposed the "3E Concepts", which is ecology, economy and equity. Three months ago, I am relieved of my office there, but for the value of the "3E Concepts", I think there is no stage or time limit and it is worthy of attention and consideration of every one of us present.

Not long ago, I learned from a material that in 2010, the urban population firstly outnumbered that in rural areas and the world is experiencing the rapid transformation from a society dominated by rural areas to that dominated by urban areas. There is no doubt that we are in an era when unprecedented urbanization is transforming the world we live in, we have to be prepared for it. One of these prominent problems is climate change. Because climate change has posed a threat to humankind's development and existence, it deserves to receive wide attention from countries of all over the world. For example, Dar es Salaam, a coastal city of Tanzania, is low-lying. It might be submerged if the sea level of nearby sea area rises by over one meter. Therefore, the foremost thing for us is to deal with climate change and work in concert to construct an eco-safe and harmonious world in which humankind and nature developed coordinately.

I got acquainted with Mr. Jiang Mingjun as early as year 2004 when I attended the World Political Forum in Italy, where he impressed me with his eco-safety concept. Later he and his team participated in three sessions of World Urban Forum held by UN-HABITAT and organized Side Event themed "Eco-Safety and Sustainable Urban Development" at each of the forums, thereby establishing a good cooperative partnership with UN-HABITAT.

One of the big challenges we are facing, however, is the contradictions between most people's intense desire of pursuing modern life and the carrying capacity of the earth's natural resources and ecological environment, and the contradictions become increasingly sharp. Though the present trend is that every one is talking of ecological protection, the action is not really in consistent with the idea.

From a global point of view, various unexpected natural calamities and ecocatastrophes almost happen every day. Every time we get to the bottom of the disaster after it happens only to find that the root of problem lies in ourselves. It is our humankind's excessive claim from nature and our unending desire that destroy the balance between human and nature. So, ladies and gentlemen, the realization of ecological safety and sustainable development hinges on primarily what kind of orientation, pattern and progress of modernization are chosen. The government should seek more actively a series of new solutions, implement positive social management and development plan and promote social equity.

Meanwhile, nongovernmental organizations and citizens are also expected to play a constructive role, for it is not possible for the government regulation to cover every aspect. I suggest that they make best of their advantage of belonging to some particular communities to put forward their suggestions and original ideas to the government and society; promote social harmony with the help of internet and forum. Thus we can jointly commit ourselves to the advancement of ecological safety and sustainable development.

Lastly, I wish the Assembly a great success!

A Global Debt-for-Environment Proposal to Finance Climate-Change Programs

Jose de Venecia Jr., Founding Chairman of the ICAPP

We are pleased to be invited to open this First World Eco-Safety Assembly that is being held in conjunction with the Sixth General Assembly of our International Conference of Asian Political Parties here in Phnom Penh, in this cool and pleasant season. We are honored to be associated with the historic efforts of IESCO, the Funcinpec Party under Deputy Prime Minister Keo Puth Reasmey, and the Cambodian People's Party under Premier Hun Sen, Deputy Premier Sok An, and the Cambodian Parliamentary leaders Heng Samrin and Chea Sim.

Environmental Protection as an ICAPP Mandate

Protection of the environment is of course among ICAPP's basic mandates and IESCO's foremost challenge. We realize the earth is not for any generation to treat as it pleases, but is held in trust for those who will come after us.

We're also deeply aware the ecological crisis has become one of the direst threats to our collective life. Our decades of wastefulness and neglect are being paid back in ecological catastrophes: by storms, floods, and droughts — the shrinking of wetlands; the spread of deserts; the extinction of species.

And, as always, it is the poorest countries that suffer the most from these 'natural' calamities that may wipe out — in the twinkling of an eye — tens of years of material gains that people may have won from patient husbandry, hard work and thrift.

To Reverse the Trend, Ecological Cultures

To reverse this downward spiral toward ecological collapse, humankind must learn to live in harmony with nature. We can no longer exist as the arrogant and heedless 'Masters of the Universe.' We must begin to rule all creation as the Almighty's just stewards — and begin to renew the face of the earth.

Fortunately, everyday people are beginning to nurture ecologically-sound cultures. Organic food has become popular all over the world, particularly in Europe. China's Communist Party has declared its goal of an 'ecological civilization' as a guiding strategy of national development. And the global community — under United Nations auspices — has begun to negotiate a strategy to moderate global warming and climate change.

Problems of the Climate-Change Talks

The toughest aspect of these negotiations is that the newly-industrializing countries (NICs) are unwilling to accept internationally-binding emission reduction targets on carbon pollutants without financial or technical compensation to cover the economic costs of achieving these targets — since the NICs have historically generated much lower per capita emissions than the industrialized countries, and will continue to do so over the foreseeable future.

Debt-For-Environment Swap for Poor Countries

To overcome this problem — which stalemated the initial negotiations in Copenhagen last year — we in the Philippines proposed a substantial 'debt-for-environment formula' to complement the debt-for-equity plan to fight poverty for the 100 highly indebted middle-income countries under the program to finance the U.N.'s Millennium Development Goals (MDG's)

Our proposal is voluntary and does not ask international creditors to forgive — or suspend — a single dollar of debt. Nor will it require new money from the legislatures of any of the rich countries: the U.S. Congress, the British Parliament, the Japanese Diet, the German Bundestag, China's National People's Congress, or the French National Assembly.

The proposal is simple. It offers creditor states and lending institutions the option of converting up to 50% or portions thereof of the debt-service payments they receive into investor equity in environmental programs in their debtor countries.

These projects would include reforestation, water conservation, alternative energy, mass housing, health, education, eco-tourism, and other social infrastructure.

Social Projects are Commercially Viable

Many of these social projects are potentially viable commercially: they can generate tolls, users' taxes, rentals, and sales income. Reforestation projects earn carbon credits under the Kyoto Protocol. Ethanol from sugar cane has proved a potent substitute for petroleum. And a cycle of timber sales can be profitable, as witness the evergreen timber plantations of Canada and New Zealand. Eco-tourism too can pay its way in jobs and profits.

Mass housing projects will enable poor-country governments to relocate millions of squatter families living along watercourses and on sloping land susceptible to landslides.

As equity-holders in these projects, creditor countries will have at least oversight powers over the projects they finance.

For the poor countries, debt-for-environment projects — reforestation in particular — will have a tremendous economic-stimulus and job-creation impact amid the global downturn. Funds diverted from debt service to foreign direct investment in environmental programs will pump-prime poor-country economies. They will generate many jobs for relatively low-skilled people.

Our Proposal Recognized by the General Assembly

The original Philippine debt-for-equity proposal — to finance anti-poverty projects and the U.N. Millennium Development Goals (MDGs) in the 100 most-heavily-indebted countries — was recognized by the General Assembly, as well as by the governments of Germany, Italy, and Spain; the Group of 77 Plus China; the International Conference of Asian Political Parties (ICAPP); and the Eleventh Summit of Presidents and Prime Ministers of ASEAN, the Association of Southeast Asian Nations.

China on Climate-Change Financing

Among the great powers, China has come down on the side of the poor countries on the issue of climate-change financing.

Replying to the Philippine proposal, Yu Qingtai, Beijing's special representative for climate-change negotiations, noted that the rich and the poor countries have "common but differentiated responsibilities" under the United Nations Framework Convention on Climate Change.

Under the Convention, the developed countries are obliged to provide financial resources to the developing countries to cover the costs of the NICs' achieving their carbon emission reduction targets.

China's declaration on the financing issue spells out what both the rich and the poor countries must do cooperatively to clean up the atmosphere.

IESCO's Gallant and Commendable Effort

In conclusion, I must say our debt-for-environment formula is only one of the many complex measures the global community must take to ensure we succeed in our campaign against environmental degradation and climate change.

And I must add that little hope is being held for a substantial agreement at the second session of the UN climate-change conference in Mexico City this December.

In this setting, this First World Eco-Assembly is a gallant — a truly commendable — effort by IESCO, and we thank our most passionate leader, the Honorable Jiang Mingjun, and the world leaders who are here, for their powerful advocacies.

You have resolved to do what you can — and we in ICAPP feel privileged to be counted with you in this your effort to help ensure the ecological integrity of our home-planet.

On ICAPP's and IESCO's behalf and our new affiliate, the Centrist Asia Pacific democrats International (CAPDI) — and on my own — I wish our Conference the success it deserves.

Compelling Politicians to Govern Green

What can we in IESCO and ICAPP do?

The campaign for environmental protection will be a long-term cause. We can become the World Eco-Safety Assembly's political spokesmen. We can muster the Asian political parties on the environment's behalf. We can form a united front against climate change.

We can begin to make ordinary people aware of the imperatives of ecological safety — and the power they have — through their collective vote — to compel those who rule them to govern green — to ensure the interests of the environment are represented in their every public policy decision.

Include Protection of Environment in Constitutions and Party Platforms

Through ICAPP, IESCO and CAPDI, we can try our utmost and perhaps see to it that ecological safety — safeguards against environmental degradation — and green public policies — are written into every national constitution — every political party's platform — and every civil society charter.

These, I believe, are the least we could try and do — to show our solidarity with the universal cause that we and IESCO and many distinguished other leaders have begun to champion.

On ICAPP's and IESCO behalf — and on my own — I wish our Conference the success it deserves.

Thank you and good day.

Protecting Ecological Environment Is Protecting Human Civilization

Mendsaikhan Enkhsaikhan, Former Prime Minister of Mongolia

Respectable Chairman of the Assembly,

Distinguished guests,

ladies and gentlemen,

Good Morning! I am very glad that I can attend the First World Eco-Safety Assembly in Phnom Penh, Cambodia. In today's world, ecological safety has become a hot topic that is known in every household. I feel greatly honored to express my opinion through this Assembly. Here, I'd like, on behalf of the Government of Mongolia, to express my warmest congratulations on the convening of the First World Eco-Safety Assembly!

The survival and development of human beings rely on nature, and affect the structure, function and evolutionary process of nature as well. The relationship between man and nature is embodied in two aspects: first, mankind's influence and effect on nature, including extracting resources and space from nature, enjoying the service function provided by ecological system, and discharging waste to environment. Second, nature's influence and counteraction on mankind, including resources and environment's restriction to the survival and development of human beings, the negative influences of natural disaster, environmental pollution and ecological regression on human beings. With the continuous development of productive forces of human society, mankind's ability to explore and utilize nature keeps increasing. Making a general survey of the development history of human society, we can see it is actually a history of blood and tears for ecological environment. In its own way, the earth is punishing human beings' predatory deeds on nature which are close to insanity: depletion of natural resources, global warming, expanding of ozone holes, serious pollution of atmosphere, water and soil, accelerated extinction of species, grievous desertification, rampant floods, and so on.

Located in the central Asia, Mongolia is a country with unique ethnic customs and natural geographical landscape as well as rich resources. There are over 80 mineral products that have been ascertained such as copper, gold, silver, uranium, lead, zinc, rare earth, iron, and fluorite.

However, with the excessive development and utilization of natural resources, Mongolia is facing various environmental pressures such as environmental resources and environmental property rights, air pollution, utilization of resources of livestock husbandry, protection and utilization of forests, location and exploitation of water resources. Among all these problems desertification is the most serious one. Mongolia has been listed by the United Nations as one of the eleven countries that are under the uttermost threat of desertification. Up till now, desertification of different levels is attacking over 72% land of Mongolia.

Conservation of forest plays the main role for combating desertification and global warming. The only and most efficient way to conserve forest is through empowerment and enabling framework for local residents who are residing in forest covered areas. Therefore, the Government of Mongolia brought an important amendment into the existing Law on Forest in 2007 which enables and empowers the local people to own a certain forest fund that they are living aside via a contractual basis. As the result of the new amendments, there are many Forest User Groups are established and protecting and sustainably using the forest resources which greatly reduce illegal logging and forest fires in their forest fund areas. These are certainly a great contribution to nature conservation and livelihood improvement at the same time.

As a matter of fact, desertification is an ecological conundrum not only for Mongolia but also for the whole world. About one third population of the world lives in the areas that are endangered by desertification. Like a plague, desertification is bothering North America, Australia, Central Asia as well as the Middle East, too. I always believe that mankind can not get rid of civilization's dependence on nature and nature's restriction on civilization no matter how hard mankind pushes forward its civilization. The decline of natural environment will certainly be the decline of human civilization. Therefore, let us make contribution together to the improvement of global ecological environment.

Thank you!

Give Full Play to International Organizations in Eco-Safety

Zhang Deguang, First Secretary-General of the Shanghai

Cooperation Organization. and Co-Chairman of IESCO

Respected Chairman of the Assembly,

Distinguished guests,

ladies and gentlemen,

Hello, everyone! Today, we are gathered here in Phnom Penh, Cambodia, to attend the Sixth General Assembly of ICAPP and the First World Eco-Safety Assembly. I would like to, on behalf of IESCO, take this opportunity to give our warmest welcome to all the distinguished guests present.

The world today is in an era of deepened economic globalization and accelerated political multi-polarization, a distinctive feature of which is that no certain power can dominate world affairs, and international cooperation and communication are required, especially in some problems concerning global development, such as the handling of climate change, maintenance of eco-safety, protection of eco-environment, promotion of urban development, reduction of poverty and implementation of energy-saving and emissions reduction.

The greatest problem our humanity is facing in the 21st century is how to jointly respond to various complicated challenges and promote global sustainable development. Therefore, climate change and ecological safety have become hot topics in current international society, and "environmental diplomacy" and "ecological diplomacy" have become new diplomatic subjects for countries all over the world.

One of the features of this Assembly is that the International Conference of Asian Political Parties is one of the sponsors, Cambodian People's Party and Funcinpec

Party play the host, and IESCO, United Nations University and China Foundation for International Studies involved themselves in sponsoring or assistance, which reflects the unique role of nongovernmental organizations in promoting global eco-safety and sustainable development in line with the trend of globalization and multi-polarization.

In recent years, many international organizations and nongovernmental organizations have done a lot in the field of eco-safety and sustainable development. They have acted as an important channel in reconciling and resolving disputes in the field of climate change and eco-safety through dialogues and cooperation at various levels.

Dr. Jiang Mingjun, Director-General of International Eco-Safety Cooperative Organization, has made important contributions to the promotion of world ecological safety, which involves at least the following three aspects:

Firstly, he founded International Eco-Safety Cooperative Organization in 2006, when the concept of ecological safety was of little awareness in the world. The foundation of International Eco-Safety Cooperative Organization reflects his farsighted diplomacy decision.

Secondly, after years of tireless research based on his rich experience, he has proposed a set of systematic theories of ecological safety, and recently published Comprehensive Introduction on Eco-safety, a book that embodies the results of his research and breaks a path for in depth scientific research in this field.

Thirdly, with a profound understanding of the trends of world development and an aim to maintain ecological safety, promote ecological conservation and create an effective platform for international cooperation, he has overcome many difficulties to successfully hold the First World Eco-Safety Assembly in Cambodia, which welcomed participants including party and government officials as well as leaders and experts of nongovernmental organizations, enterprises, and art Institutes from dozens of countries, thus creating a new model of international cooperation in ecological safety.

The maintenance of ecological safety is the common responsibility of every country, every social class, all walks of life, and every citizen of the globe. It is my sincere hope that more people will emerge from the international community to have an enthusiastic commitment to the cause of ecological safety, like Mr. Jiang Mingjun. I believe that the convening of World Eco-Safety Assembly can set up a broad and effective platform that attracts more national governments, parties and parliaments, international organizations, nongovernmental organizations, financial institutions and enterprise groups to play a positive role in promoting global ecological conservation, addressing climate change, maintaining eco-safety, protecting eco-environment and realizing United Nations Millennium Development Goals in joint efforts.

Thank you!

II

Keynote Speeches

(Climate Change and Ecological Safety)

Actively Cope with Challenges Posed by Climate Change

Mok Mareth, Minister for the Environment of the Kingdom of Cambodia, and Chairman of the National Climate Change Committee

His Excellency Chairperson,
His Excellency Director - General Jiang Mingjun,
Distinguished National and International Guests,

It is a great pleasure and privilege for me to deliver the welcome remarks for this First Annual Meeting of World Eco-Safety Assembly today on behalf of the Royal Government of Cambodia. I wish to extend my warm regards and welcome to you all to Phnom Penh – the Capital of Cambodia. I realize that you are fully dedicated to the sessions as mentioned in the agenda, but I do also hope you will take time to enjoy a fascinating Cambodia, especially a glorious land of Siem Reap where Angkor Wat Temple/Complex – the world cultural heritage is located.

Excellencies, Ladies and Gentlemen, Cambodia emerged from the 3-decade civil war just about ten years ago. As a post-conflict society, we are facing many priorities and challenges in rehabilitating and developing our economy, infrastructure and social services to improve people's livelihood, eradicate poverty, and to ensure sustainable development in accordance with the Rectangular Strategy Phase 2 of the Royal Government and the National Strategic Development Plan Update 2009-2013. The Royal Government of Cambodia has made significant progress in developing infrastructure and irrigation system to support the agricultural sector development as the backbone of the country economy and a key toward poverty alleviation. An ambitious goal has been set up to make Cambodia a net rice exporter with a target of

1 million tons by the year 2015. Fisheries reform is underway to ensure that fish will remain a vital source of protein for the whole population.

A comprehensive environmental legal framework has been in place to support rational use of our natural resources and high environmental quality. Our conservation effort is being realized for protected areas and other forest conservation areas, which cover about 25% of the country's total land area. We have an ambitious goal of increasing the forest area up to 60% of the total land area by 2015. We are now working to nominate our 435 km coastal area into the "Club of the Most Beautiful Bays in the World". We are really proud that our efforts following the concept of "balancing development and conservation, conservation to support development" have resulted in reducing poverty from 50% in 1990 to 27% now while at the same time we have managed to maintain high environmental quality. This also leads to reducing pressures on the natural resources.

Overall, Cambodia has enjoyed social and political stability and steady economic growth, thanks to the able leadership of Samdech Akka Moha Sena Padei Decho Hun Sen Prime Minister of the Kingdom of Cambodia and Honorary Chair of the National Climate Change Committee. We also started to play an increasing role in the regional and international arenas. Cambodia has proposed a number of key regional initiatives to address regional and global challenges and to promote cooperation. We have sent our peace keeping forces to many countries to support our humanitarian missions.

We are doing this in an era where climate change has emerged as a key factor at the national, regional and global levels. As a country of the Great Mekong Sub-region and as an agrarian economy, Cambodia will be severely affected by impacts of climate change. More scientific evidences have emerged showing that the Mekong will be severely affected by climate change. The effects of climate change on water flow, fisheries, wetland and watershed functions will lead to serious negative impacts on people livelihood and may undo the hard-gained development achievement.

Last year Cambodia was hit by a very strong typhoon Ketsana which caused considerable casualties and damages to crops, infrastructure, and households. This year, we have experienced unusual delay of rainy season and extreme rainfalls. Just about a month ago, an extremely high rainfall of over 50 cm from a single rain was recorded in Takhmau city near Phnom Penh causing a flash flood over a large area.

The Royal Government of Cambodia under the leadership of Samdech Decho Prime Minister Hun Sen clearly understands this new challenge and as a party to the

UN Framework Convention on Climate Change (UNFCCC) and its Kyoto Protocol, Cambodia has been making its outmost efforts to fulfill its commitments under these international treaties as well as to take measures to respond to climate change within the county. Recently, Cambodia participated in the 15th Conference of the Parties to the UNFCCC with clear positions related to future commitments to reduce greenhouse gas emission, needs for substantial financial and technical support for vulnerable countries to adapt to climate change, and for technology transfer to developing countries. We stressed the importance and urgency of adaptation to climate change for developing countries. We urged developed countries to honor their commitments under the UNFCCC to significantly cut their emissions and meaningfully provide financial support to developing countries. We have consistently maintained these positions along with G77/China Group.

We strongly believe that to achieve the ultimate objective of the UNFCCC, atmospheric concentration of greenhouse gases should not exceed 450 part per mission so that temperature increase will not exceed $2^{\circ}C$. Therefore, aggressive emission reduction should be implemented by all countries based on the "Common but Differentiated Responsibilities" principle of the UNFCCC. In this regard, we are very encouraged by progress in GHG emission reduction by some developing countries, including China which is becoming a leading country in renewable energy development and energy efficiency.

At the regional level, we have been active in promoting regional environmental and climate change cooperation with ASEAN, GMS and Mekong River Commission in accordance with the Mekong Agreement – 1995 focusing on the cooperation in all fields of sustainable development, utilization, management and conservation of the water and related resources of the Mekong River Basin. We urge careful and comprehensive social and environmental impact assessment, as well as cost-benefit assessment, of any development projects on the Mekong tributaries and along the Mekong both upstream and downstream, partnership, dialogue and information sharing to promote equitable, well-balanced and sustainable use of the vital shared resources of the Mekong. In this regard, Cambodia strongly supports the development and implementation of a Basin Development Plan (BDP) to promote, support, cooperate and coordinate in the development of the full potential of sustainable benefits to all riparian states and the prevention of wasteful use of the Mekong Basin waters. We strongly support continuous implementation of the Water Use Programme to improve shared water use and management in the basin while promoting appreciation of the importance of ecological balance. An all of this have to be done in the context of climate change and the need for mainstreaming adaptation into all development policies, plans and activities.

Excellencies, Ladies and Gentlemen, the Royal Government of Cambodia under the able leadership of Samdech Decho Prime Minister Hun Sen, vice Chair of the Cambodian People Party, is committed to do its part to address the above-mentioned global challenges. Based on the existing national policy and legislation, regional and international agreements, we are committed to work with all countries to combat climate change, reverse vital resource degradation and to preserve our planet for our children and children of our children. Regionally, we have full political will to make Mekong a model of regional cooperation, a symbol of peaceful coexistence and mutual prosperity.

We the human species have been taught to believe that we have the ultimate power over the mother earth to alter her to serve our purpose, needs and often greed. The recently emerging crisis of climate change has once again reminded us that nature has its limit and we simply cannot continue our behavior as we have been doing toward her if we are about to survive as a species. We may already reach the point of no return where climate change and other environmental challenges start to alter the path of our civilization. The urgency is now and I do hope that this First World Eco-Safety Assembly will serve as a catalyst for a global concerted effort to save our planet and ultimately our species. I am confident that during this one-day event, we all will enjoy a spirit of solidarity and shared responsibility toward our only habitable planet, mutual understanding, rich contribution from you all and promising outcomes.

I wish you a very successful Assembly and thank you for your attention.

Ecological Safety: Foundation of Peace and Development

Jiang Mingjun, Director-General of IESCO

Respected Y.E. Prime Minister Mendsaikhan Enkhsaikhan, Dr. Anna Tibaijuka, UN Under-Secretary-General, Y.E. Under-Secretary-General Anwarul K. Chowdhury, Y.E. Deputy Prime Minister Keo Puth Reasmey, Minister Dr. Mok Mareth, Minister Zhang Deguang, Minister Yang Shen, honorable guests, ladies, and friends,

First, I'd like to express my thanks to the Royal Government of the Kingdom of Cambodia, ICAPP, the Cambodian People's Party, the Funcinpec Party, United Nations University, China Foundation for International Studies and other organizations for their concern for and support in this Assembly, and also express my gratitude to leaders of political parties, national political figures, experts and scholars for their presence at the Assembly.

Next, please allow me to make a work report on behalf of International Eco-Safety Cooperative Organization. And now please examine my work report.

I. The present condition of ecological crises and countermeasures

As we all know, recent years have witnessed the global natural disasters and severe ecological disasters one after another, such as earthquake, tsunami, hurricane, volcanic eruption, flood, drought, mud-slide, forest fires, land desertification, sandstorm, water resources contamination, environmental contamination, industrial contamination, sharp decline of wetlands, species extinction; destruction of marine ecosystem caused by crude oil spill, geological disasters caused by unscientific water conservancy projects and overexploitation of underground water resources; epidemic diseases such as Aids, SARS, bird flu, mad cow disease, foot and mouth disease, cholera. Those ecological disasters caused by climate change and human economic activities have posed a serious threat to human survival and national development. Moreover, the frequency of disasters, causalities, and economic loss are still increasing. It is beyond doubt that a local war requires a long diplomatic process. However, an unexpected ecological disaster, which is instantaneous, can create far more casualties and economic loss than a local war.

As early as 1998 when I was the Secretary-General of International Coordinating Committee for Asia and the Pacific, I had broke the theoretical shackles and put forward for the first time the concept that "the biggest political issues in the 21st century are ecological safety and resources security", because the disputes and conflicts that arise between countries and between regions are directly linked to these two aspects. Disputes over water resources in Sudan's Darfur region and among India, Pakistan and Bangladesh are sufficient to prove this point. My concept immediately aroused high attention from countries such as Russia, America, United Nations, European Union and international organizations after it being put forward. In early 1999, in Russia the National Eco-Safety Commission was founded, which is directly under the leadership of Russian Federal State Security Commission. At the end of the same year, America established the state eco-safety administrative institution. In recent years, though America pays close attention to domestic issues in eco-safety research, it attaches greater importance to global ones. They think that worldwide ecological crises have threatened the prosperity of America. But they also realize that America may get involved in regional conflicts and conflicts between countries caused by ecological crises, pay heavy price for them and even launch dangerous military intervention.

II. Reflection on the First Five-Year and the Second Five-Year Plan

1. Reflection on the First Five-Year

In February 2006, in face of the frequent ecological crises, planned by China, with the support and participation of related UN agencies, International Eco-Safety Cooperative Organization was founded in Hong Kong, China, which aims to maintain ecological safety and cope with ecological crises through cooperation with political parties, parliaments, non-governmental organizations and enterprise groups, and regards the realization of UN Millennium Development Goals as it's duty. Since it was established, we have carried out fruitful work centering around international conference, project cooperation, and scientific research: First, we held the First City Diplomacy Conference in Hague, Netherlands, held the conference Peace With Water with European Parliament in Brussels, attended the Second, Third, Fourth, Fifth World Urban Forum that hosted by UN-HABITAT and governments of UN member states in Barcelona, Spain, Vancouver, Canada, Nanjing, China, Rio de Janeiro, Brazil respectively. We hosted thematic session and exhibition on Ecological Safety and Sustainable Urbanization, and co-hosted the Forum on Food Safety and Ecological Safety with United Nations University, China Foundation for International Studies and PA International Foundation in Beijing, China. Over the five years, we entirely hosted and participated in 37 important international conferences, and signed 6 strategic documents such as World Urban Diplomatic Declaration, World Convention on Peace with Water and Joint Declaration on International Ecological Safety. Second, we established three international eco-safety demonstrative cities, four best paradigms of new countryside construction, five international ecological tourism demonstrative spots. The balanced development among economy, ecology and society is embodied through the practice of the above-mentioned projects. Third, we actively promoted eco-safety technologies. The High-energy Fuel, a microorganism ecological energy which is fermented and synthesized by industrial sewage, seeping fluid of waste, domestic sewage and special microbial community and developed by researcher Deng Chubo, fills the vacuum in the field of world ecological energy, and attracts attention and concern from countries such as the United Nations and European Union. The Taiji "slag desulfurization + saline-alkali land improvement" technology researched by researcher Shi Hanxiang can be used to treat saline-alkali land and sandy wasteland. It can return to us both the blue sky and the green land, thus is the best paradigm of using waste to treat waste and realization of comprehensive utilization of resources. The severe saline-alkali land improvement technology, which was researched by researcher Zhu Qijiang, can transform a large amount of severe saline-alkali land into fertile farmland. The Juncao planting technology developed by Academician Lin Zhanxi is applied in countries such as South Africa, Lesotho and Rwanda to carry out poverty alleviation. It not only solves the employment of local people and reduces poverty but also attracts attention and support from the governments of the countries where it is applied. Plant lactic acid bacteria technology, the research Academician Wang Houde was in charge of, is applied to the development of ecological agriculture and plays a significant part in ecological planting and cultivation. The architectural anti-earthquake technology, which was researched by Academician Peng Peigeng, is so far one of the safer, more applicable and economical anti-earthquake technologies, and has been adopted by UN. The "acid membrane" planting technology, which was researched by Academician Wang Jiayou, not only plays a favorable role in preventing sand and treating desertification, but also attracts concern and attention from Chinese leaders and related departments.

Over the five years, we, together with Chinese Ministry of Environmental Protection and State Bureau of Forestry, hosted the first All China Safari to promote ecological and environmental protection. This safari covered 22,000 kilometers and encircled almost half of China. During this safari, we investigated and surveyed the condition of air pollution, contamination of water resources, as well as the discharge of waster water, waste gas and waste oil in cities and provinces along the safari, provided timely advice for local governmental institutions and assisted them in formulating improvement measures. Moreover, we also actively participated in the relief and post-disaster reconstruction after the 5•12 Wenchuan earthquake and Haiti earthquake, published 27 issues of International Ecology and Safety and 9 series books on international ecology and safety.

In recent years, we also provide capital for the education, poverty alleviation and subject study of the less developed countries such as China, Tanzania and Nepal as well as poverty-stricken regions. At present, we have forged strategic cooperative organization with 83 national and international organizations and have finished the work of the first Five-year Plan as scheduled and laid solid foundation for the second Five-year Plan.

2. The Second Five-year Plan

Seizing the 6th General Assembly of the ICAPP and the 1st Word Eco-Safety Assembly as an opportunity, taking Phnom Penh Declaration and (Angkor) Resolution of World Eco-Safety Assembly as guidelines, we will make a detailed plan of the Second Five-year development strategy and lay out the Second Five-year plan.

First, pay adequate attention to the four main conferences, that is, preparatory work of World Eco-Safety Assembly and First World Eco-Safety Expo which are to be held in different countries biennially, Climate Change and Eco-Safety Forum which is to be held in China in June, 2011, with United Nations University and China Foundation for International Studies, International Painting and Calligraphy Competition and itinerant exhibition featuring ecological safety and natural disaster.

Second, cooperate with agencies of United Nations and European Commission, keep carrying out the guidance of construction of eco-safety city, community, park, tourist area and enterprise, and executing evaluation and certificate of eco-safety products (projects).

Third, continuously carry out poverty reduction program through offering technology and training in the less developed countries and the least developed countries in Asia and Africa to cope with the disasters and influence brought by climate change in the above-mentioned countries.

Fourth, actively support United Nations University in conducting subject research in the fields of climate change and ecological safety, participate in preparatory work of climate change and ecological safety college of United Nations University, provide strategic advice to United Nations, the ICAPP, political parties, parliaments and governments of the various countries in fields of climate change, ecological safety and sustainable development.

Fifth, do good job of the publishing of International Ecology and Safety, operating of the website of World Eco-Safety Assembly and our organization website in a good way, further strengthening theoretic study on ecological safety, carrying out academic exchange and international multilateral cooperation, and issuing of the annual report on ecological safety.

Sixth, do good job of organizational and institutional construction and consolidation of monitoring mechanism so as to ensure the scientific management and orderly operation of work of our organization. Seek support from governments of various countries and international organizations, and establish "Headquarters of International Eco-Safety Cooperative Organization".

III. A few pieces of advice and countermeasures

First, after the world political pattern enters the multipolar period, no single power can dominate the world single-handedly, especially the major issues that concern global development, such as climate change, deforestation, loss of biological diversity, desertification, contamination of water resources, frequent occurrence of geological disasters. These issues can not be controlled by certain country or region alone. Urgent tasks before every country in the world at the moment are how to stop the rapid depletion of natural resources on the earth, how to keep the balance among economic, social and ecological development for both the present and future generations, and how to strengthen the cooperative network among the developed countries and the developing ones. We can no longer seek temporary economic prosperity at the cost of ecological environment. In face of the frequent natural disasters and ecological catastrophes, we must establish multinational and cross-border joint network for early warning instead of waiting passively for our end. First, we should deepen our understanding of the concept of ecological safety and environmental protection. The various countries in the world should bring ecological safety and environmental protection into national educational system, and learn carefully from New Zealand and Japan's experience in earthquake prevention and disaster reduction as well as Chile's disaster prevention and rescue mechanism. Second, national leadership and coordination institutions on eco-safety should be set up to wholly prevent the various departments from shifting the responsibility onto others and bureaucratic corruption in the aspects of climate change, ecological safety and disaster aid. Third, efforts should be made to strengthen legislation on ecological safety and environment. The Republic of Seychelles and the Kingdom of Cambodia have started off before other countries. They all define ecological safety and environmental protection as the foundation to build their countries, and carry out legislation on ecology and environment. The practices have proved that only when ecological safety and environmental protection are brought into legal system can ecological crises be effectively curbed, and balanced development among ecology, economy and society is realized.

Second, it is imperative to seek common ground while reserving differences, strengthen international multilateral cooperation and jointly cope with disasters and conflicts caused by climate change and human activities. By and large, differences in cultural background and level of development between different countries in the world, which are extremely obvious, could pose some difficult problems for the cooperation of various countries in the world. However, as long as every party acts on the principle of "seeking common ground while reserving differences, and pursuing a harmonious win-win situation", the construction of a new pattern which promotes peace and development will be greatly facilitated. The present overall situation of ecological crises features local improvement and overall deterioration. Therefore, under the general framework of UN Millennium Development Goals, every country should carry out the basic principle of "Common and Differential Liability". The developed countries should actively and positively bear the responsibility of the maintenance of eco-safety and environmental protection, and play more significant role in the realization of peace and development, whereas the developing countries should combine ecological construction with economic development, enhance their capacity of sustainable development, and positively tackle climate change and participate in international cooperation and competition.

Third, ecological disasters may break out at anytime under the current circumstances of mass migration and rapid urbanization. Therefore, it is necessary to build a disaster emergency response mechanism and set up permanent rescue and aid institutions with aims to raise the governments, society and the public' awareness of ecological disasters and build up the self capability to deal with emergency rescue and aid by various effective means. We support the ICAPP, Malaysia, Japan and other countries in establishing International Emergency Rescue Forum of Asia and International Emergency Rescue Center of Asia and setting up World Eco-Safety Fund (tree-planting). The UN Millennium Development Goals is focusing on global cooperation and development in an unprecedented active way, and IESCO is ready to make new and significant contributions to achieving these goals.

Preserve Regional Eco-Safety
— Concept and Practice

Li Wenhua, Academician of Chinese Academy of Engineering

I. Service function of ecology is the important guarantee of eco-safety

Eco-system is the most basic unit and the most active part of biosphere. It not only provides the humanity with various goods, including food, medicine and raw materials for industrial and agricultural production etc., but more importantly, supports and maintains the earth's life system, including soil and water conservation, carbon fixation and oxygen release, nutrient substance accumulation, environment purification, bio-diversity conservation, recreation and aesthetic value etc. Increasingly, people are realizing that the service function of ecological system is the foundation of humanity's survival and development. According to the study of Costanza etc. in 1997, the global eco-system creates a value of about 33,000 billion dollars per year, 1.8 times more than that of global GNP. Though there is controversy about the accuracy and method of calculation, this study played an important role in arousing people's regard for the service function of eco-system, and became the core of the Millennium Eco-system Assessment (MA) initiated by United Nations in 2000.

II. Challenges and opportunities that face the eco-system in China

China's natural environment is congenitally deficient. The arid and semiarid regions cover 52% of the country's total land area, mountainous land and very rough terrains make up two-thirds of its territory. The Tibetan Plateau covers an area of 2 million square kilometers, karst region 0.9 million square kilometers and Loess Plateau 0.64 million square kilometers. The vulnerable natural conditions cannot sustain the large population and bear the pressure brought about by rapid economic growth.

Since the reform and opening-up, the rapid economic development has led to a concentrated reflection and outbreak of ecological problems, which appeared in different development stages in western developed countries, in a short term. Though, in recent years, the ecological conditions in China have been improved, the eco-system degradation is still in progress, which has become a bottleneck in the sustainable development of economy and society, as well as will have a far-reaching influence on people's health, economic development, social stability and even national security.

According to relevant report, the area of water loss and soil erosion in China has reached 3.56 million square kilometers, occupying around 37% of the country's land area. The area of desertification is about 1 million square kilometers, expanding at a speed of 3,436 square kilometers per year. While the area of woods has increased, covering an area of 0.195 billion hectares, 20.36% of the land area, it is still lower than the world average of 29.6%. Moreover, the forest quality has gone off and the structure of age of stand is unreasonable, which have led to a continuous reduction of recoverable resources. Ninety percent of grasslands have degraded in various degrees, with half of them gently degraded, and the grassland area of "Three Degradation" in the country has reached 0.135 billion hectares, with an increase of 2 million hectares per year. Groundwater over-exploitation is so serious that there appeared in North China Plain the greatest underground complex funnel area in the world, stretching forty or fifty thousand square kilometers, and, vast stands of trees died for lack of water in many places of west area. Bio-diversity has reduced sharply, displaying by the reality that in the 1,121 endangered species listed in Convention on International Trade in Endangered Species of Wild Fauna and Flora, out of which there are 190 species in China. In addition, the invasion of foreign species has posed a threat to eco-safety and caused great economic losses. The environmental pollution has spread from land to coastal waters, stretched from surface water to groundwater and extended from common pollutants to toxic and harmful pollutants, forming a situation that point source pollution and area-source pollution coexist, domestic pollution and industrial discharge mix and various old and new pollution and secondary pollution combine.

III. Efforts and achievements which China has made in ecological protection

On the other hand, it should be noted that the Chinese government and people have made great efforts in ecological construction. Tremendous human, physical and financial resources have been invested to ensure the ecological safety of the country's land and remarkable achievements have been made. The following aspects are especially worth getting our attention:

Natural forests protection project. The project covers 18 provinces and autonomous regions. Measures of banning and limitation of lumbering in natural forests have been taken in an effort to recover and restore the forests' production and development. By implementing this project, the timber yield was reduced 19.91 million cubic meters, afforestation area reached 0.013 billion hectares, 0.1 billion hectares of forest area were managed and conserved, and 740,000 surplus labor force were reassigned.

Project to return farmland to forest. The project covers 25 provinces, autonomous regions, cities and the Xinjiang Production and Construction Corps, including altogether 1,897 cities, districts and counties. Farmers were provided with free cereals and plantlets by the government in an effort to control an area of 230 million hectares of water loss and soil erosion and an area of 270 million hectares of strong winds and drifting sand by 2010.

Beijing-Tianjin sandstorm control project. The project covers 75 counties of 5 provinces, autonomous regions and cities. Mainly in an effort to solve the sandstorm problem in Beijing and its surrounding regions, it is planned that the covering of forestry and grassland will be increased from the present 6.7% to 21.4% by 2010, through measures of forests and grassland nurturing, sands-checking, forest planting, comprehensive harness of small watershed, reasonable development and application of water resources etc.

Project of construction of shelterbelts in north, northeast, northwest China and in Yangtze River Valley. The project covers 1,696 counties of 28 provinces, autonomous regions and cities. It is planned that the forest planting will reach 0.023 billion hectares and forest managed and conserved will reach 0.072 billion hectares.

Protection of Bio-diversity. China now has initially established the policies, system of laws and regulations and administrative management for bio-diversity protection. Meanwhile, several large-scale national and regional field work were organized in succession, the survey and cataloguing of species resources were strengthened, and relevant data bases of bio-diversity were organized and developed. In line with the principle and method of "in-situ conservation supplemented by ex-situ conservation", by the end of the year 2008, China had established 2,538 natural reserves (not including those in Hong Kong, Macao and Taiwan), an area of 148.943 million hectares altogether , occupying 15.13% of the country's land area and exceeding the world average of 12%, and a rationally distributed national natural reserves network of different types and complete function has initially taken shape.

Wetland conservation and construction project. A big progress has been made in the country's legislation of wetland and the draft of Wetland Protection Regulation's of People's Republic of China has been worked out. By the end of the year 2008, China had had over 550 wetland natural reserves, 36 wetlands of international significance with an area of 3.81 million hectares, 80 wetland parks with an area of 0.596 million hectares and 2.21 million hectares demonstration wetland.

Fast-Growing and High-Yielding Timber Base Construction Program in Key Areas. It is planned that, by the way of marketing and industrialization, 130 million cubic meters of timbers are going to be provided per year mainly in South China where the natural conditions are good. Currently, a new construction pattern in which diversified economic sectors take part in, various operating mechanism co-exist and develop in pluralism has been formed in the fast-growing and high-yielding timber base.

Comprehensive treatment of stony desertification program. Pilot program for tackling stony desertification in a comprehensive way was firstly started in 100 counties in 2008. The program involved 451 cities, counties and districts of 8 provinces, autonomous regions and cities in middle-south and southwest China. It covered a total land area of 1,054,500 square kilometers among which the karst area was 449,900 square kilometers and the stony desertification area was 129,600 square kilometers. The task was to construct a forest and grass vegetation of 9,420,000 hectares, to construct and transform a sloping land of 770,000 hectares, and to build the infrastructure for animal husbandry. For this program, the budgetary investment of the central authorities is 0.4 billion Yuan and the supporting capital contributed by local authorities is 810 million Yuan.

Comprehensive scientific investigation of water loss, soil erosion and ecological safety. In 2005, the Ministry of Water Resources formally launched the scientific investigation of water loss and soil erosion in which 680 people take part and which involves 292 counties of 25 provinces. The mission divided the country into 7 areas, including black soil zone of northeastern China, rocky mountain area in north China, loessal area in northwest China, stony desertification area in southwest China, the farming-pastoral ecotone in north China, upper reaches of the Yangtze River and river basins of southwest China, and red soil region of South China. The participators of the investigation covered a distance of 130,000 kilometers, held over 260 symposiums of various levels, collected 1,600 samples, made surveys at 3,200 rural households, gathered 2,100 pieces of information, took 50,000 photos and made a video of 85 hours. All of these efforts laid a solid foundation for finding out about the country's water loss and soil erosion condition and drawing up strategy of water and soil conservation.

Construction of regional ecological system. Since 1990s of 20th century, a new wave of ecological construction, mainly at the county, city and province level, has been sweeping over China. Under the guidance of scientific outlook on development and based on different levels of region units, the regional ecological construction in China has made an enthusiastic exploration in conserving ecological environment, developing ecological economy, establishing ecological civilization and constructing a new type of society at different regional levels, played a positive role in the country's sustainable development of economy and society and had good repercussions in the international community.

IV. Prospects

Although the above-mentioned programs have achieved much, there is still a long way to go for China in improving environment, conserving and nurturing ecological system due to its vulnerability of ecological environment and heavy ecological debt in history. Especially in recent years, Chinese government has put forward the people-oriented, overall, coordinated and sustainable scientific outlook on development and made solemn promise to international communication in the aspect of afforestation, energy-saving and emission-reduction. So, in order to complete these tasks, it is fundamental and critical to strengthen the nurture of ecological system, and priority should be given to the following work:

1. Administer in the light of local natural, social and economic conditions, define the principles of priority, emphasis, limitation and prohibition in the development,

and classified regulation is to be carried out in line with different types of ecological systems.

2. Maintain the continuity of the effective programs of protecting and nurturing ecology so as to strengthen the conservation and nurture of ecological systems, and promote the restoration of degraded ecological systems. In addition, continue those ecology conservation and nurture programs which have been launched.

3. Improve technology, introduce talents and increase investment. Besides, intensify publication and education, carry forward the ecological culture, mobilize the whole society, develop the recycling economy and reduce the consumption of living resources.

4. Perfect relevant polices, laws and decrees, and gradually realize and strengthen the exploration on relevant policies, especially on system construction of ecological compensation and advancing the forest tenure reform.

5. Promote scientific study on service function of ecological system.

Strengthen international cooperation and communication. As the influence of ecological system is beyond national boundaries, joint efforts of countries in protecting and nurturing the ecological systems are required and the communication and cooperation between neighboring countries are especially important. I'd like to take this opportunity to call on East Asian countries to strengthen cooperation in ecological conservation and construction in order to make the regional ecological conditions better.

The Impact of Human Activities on Tropical Diseases and Dysentery

Pierre Lutgen, Chairman of Luxembourg International

Certification Center

Climate and cultural behavior have always had a major impact on vector-borne and water-borne diseases. Vectors need specific temperature ranges for their development. In tropical regions climatic conditions provide lush environments in which innumerable hosts of infectious diseases thrive. Rainfall also plays a major role and disease vectors, like mosquitoes, are more abundant during the wet season.

As experts in the field of environmental and health certification (ISO 14000 and OHSAS 18000), we have in our international tasks been confronted by the tremendous burden of tropical diseases and dysentery. If it is true that bad environmental practices have an impact on health and diseases, the contrary is even more true: diseases and poverty have a disastrous effect on the environment, on water pollution, on deforestation, on the inappropriate use of energy, on landslides.

Diseases fall into two relatively distinct groups, the well-funded and the less well-funded like leishmaniasis, malaria, cholera, dengue, chagas, filariasis, dracunculiasis, amoebiasis, fasciola, trachoma, rabies, buruli ulcer, giardiasis… Funds currently allocated to neglected diseases are dramatically low compared to the health care costs for more common diseases .These neglected diseases are major causes of death, disability, social and economic disruption for millions of people. The lost productivity, missed educational opportunities and high health-care costs caused by infectious diseases heavily impact families and communities.

Over 9.5 million people die each year due to infectious diseases – nearly all live in developing countries. Children are particularly vulnerable to infectious diseases.

Pneumonia, diarrhea and malaria are leading causes of death among children under age 5; cerebral malaria can cause permanent mental impairment.

In partnership with associations in a dozen countries and universities in Africa and South America we have concentrated our efforts on the fight against malaria and dysentery. Our major sponsors are the ArcelorMittal Foundation and the Rotary Luxembourg Vallées.

Malaria, contrary to many beliefs, has also been a common disease in temperate countries, even in Siberia. After World War II a major effort was undertaken to eradicate this disease, with dramatic success in most Northern countries; Sicilia, Spain, Florida, Cuba, Russia, Ontario, Algeria, Greece...

The ideal tool for this was DDT, but abuses in its use lead to concerns for environmental associations and the product was banned before complete eradication was achieved in poor Southern countries. An overreaction with dramatic consequences because despite hundreds of medical studies trying to demonstrate an eventual human toxicity of DDT, so far none has been able to document any negative health effects and in Sept 2006 the WHO has lifted the ban on DDT for IRS (indoor residual spraying). In fact DDT acts on mosquitoes more by its repulsive effect than by its toxicity. IRS is a light of hope for millions of children who dye of malaria.

Although welcoming the utilization of DDT-IRS which can immediately save thousands of lives, we believe that the future lies in herbal products more than in chemical insecticides. The extract of Neem (azadirachta indica) for example has strong repulsive and insecticidal properties.

But beyond the preventive strategies based on bed-nets and insecticides, therapies based on herbal medicine can be very effective and will gain in importance. 70% of the world population still relies on these. One plant plays a key role in this novel approach: Artemisia annua. Over the last few years we have accumulated scientific evidence which shows that if tea from this plant is taken during seven days (50 gr in 20 cups) the malaria infection is completely cured with a minor risk of recrudescence. If taken longer than 7 days it can even reduce gametocytemia and transmission from man to mosquito. The Chinese know this plant for more than 2000 years and nowhere any sign of resistance to the therapy with this herb has been noticed. Artemisia combined therapy (ACT) pills however have given over the last year alarming signs of resistance. Probably because they lack the synergistic effect of polyphenols present in the dried herb.

In order to better coordinate our efforts we have launched this year BELHERB (Association for the promotion of herbal medicine). It relies on the work of a dozen university professors and medical doctors from Belgium and Luxembourg. Chinese and Indian herbal medicines have made miracles for centuries. With modern spectroscopical tools we want to better understand their pharmacokinetics.

In our own research work at Luxembourg we discovered that Artemisia annua tea has a strong sterilizing effect on contaminated water. In fact one cup of tea added to a liter of river water gives perfect drinking water. This effect has been confirmed by several European universities, but also by the universities in Senegal, Central Africa and Colombia.

The University of Antioquia in Colombia confirmed that artemisia annua has good therapeutic properties against leishmaniosis and fasciola hepatica.

The University of Belgrade confirmed that artemisia annua tea has cytotoxicity against selected malignant cell lines: human cervix adenocarcinoma HeLa, human malignant melanoma Fem-x and BG, human myelogenous leukemia K562, human breast adenocarcinoma MDA-MB-361 and human colon carcinoma LS174.

In the field of malaria, the most encouraging results were obtained by our African partners at the universities of Dakar, Bangangte, Bangui and Yaoundé. Their results are available on request.

We have started plantations in a dozen countries in Africa and South America and reach production levels for the commercialization of phytopharmaceutical products in several of these.

The scientists of "Belherb" are well aware that additional research work is required to confirm and extend above findings; in cooperation with our academic partners in the South. Should all this be confirmed, it would be groundbreaking. Each day 20000 children die of Malaria, Cholera, Diarrhea, Leishmaniosia, ...

In this period of climate changes which have dramatic effects on the health of people in Southern countries, Artemisia annua and other "Chinese" herbs could be a free medicine for them and the end of unspeakable sufferings.

Developing Green Economy Requires Pushing with All Might the Green Standardization

Zhang Bingsheng, Mayor of Taiyuan City, Shanxi Province, China

Respected Chairperson of the Assembly,
Distinguished guests,
Ladies and gentlemen,

It is my great honor to attend this Assembly and communicate with all of you. In my opinion, to preserve world ecological safety, not only relentless efforts should be made in the aspects of law, policy, technique and strategy, but a green revolution in the aspect of conception as well as the economic theory is more needed.

Point one: Push humanity transformation from "economic person" to "ecological person" in preserving ecological safety.

The replacement of agricultural civilization with industrial civilization is a historical progress in the development of human civilization. In the era of industrial civilization, human's characteristic of "economic person" was brought into full play. The hypothesis of "economic person" whose motive is the pursuit of maximization of one's own interests was firstly put forward by Adam Smith in his book The Wealth of the Nations in 1776. We should affirm that it is reasonable for human to pursue proper and necessary economic and material interests. But it is a pity that this characteristic is carried to the extreme in the real world, with the disregard for ecological interests and the excessive pursuit of material interests and comforts, all these have at last led to a serious damage to natural ecology and so on, which are the survival foundation of human itself. In order to preserve ecological safety, we must accelerate the development of human itself, discard and transcend the "economic person" and become "ecological person", referring to those who have ecological sense, emphasize the pursuit of comprehensive benefits of ecology, economy and society on the premise of respect for ecological rules consciously. With the advancement of human society, "ecological person" has been in accelerated formation as an entirely new characteristic since the 1970s of 20th century, and now the "economic person" is transforming rapidly to the "ecological person". In reality, "ecological person" is embodied by major trends such as movement of environmental protection volunteers, green movement, green international trade, green economy and waves of sustainable development, and development of ecological civilization etc., thereby enhancing the global eco-safety. It will have a fundamental effect on preserving ecological safety by accelerating the transformation of humanity from "economic person" to "ecological person", nurturing new main body of the society and realizing self-revolution, transition and enhancement on human dimension.

Point two: Push the capitalization of ecological elements and the ecologicalization of capital element and develop with all the might the green economy in preserving ecological safety.

One of the important reasons for serious ecological crisis caused by traditional industrialized pattern is that ecological resources are only used as an inexhaustible object of development and exploitation, but not the internal capital elements of economic system; that only material capital is emphasized but the value and action of ecological elements are disregarded, and the ecological elements are excluded from the capital category. Namely, the ecological system and elements are "externalized".

In order to resolve this practical question, the crucial point is to overcome fundamentally the opposition between material capital and ecological capital, economic system and ecological system, and to realize the internal integration of both. The capitalization of ecology means that as the foundation of the whole economic system, the ecological system participates fundamentally in the process of value creation, brings about directly more economic benefits and thus obtains its important property of the capital. Meanwhile, as a type of basic capital, ecological capital must realize its value preservation and increase on which human's all economic and social activities are relied and by which all of them abide. Ecologicalization of capital means that the material capital and the operation of the whole economic system must respect and embody the value and rules of ecology, especially the basic requirement of ecological capital's value preservation and increase. In reality the capitalization of ecology and the ecologicalization of capital are embodied specifically by the ecologicalization reform of traditional industrial economy and civilization, promotion of the green economy constituted by recycling economy, low carbon technology, cleaner production and green consumption etc., and the realization of transformation from traditional industrial economy and civilization to the green economy and ecological conservation.

During the past several decades, human society has been accelerating this transformation and revolution, and new roads have been opened in many places to enhance the eco-safety level, thus the eco-safety situation in some areas has taken a favorable turn. In preserving global ecological safety, national government of every country should play the role of leading force and advance with all strength the green economy as national strategy. At present, the Chinese government is advancing with all possible efforts the scientific development and quickening the pace of transformation of the mode of economic development. And the two levels of local governments where I belong to are also pushing the green economic and social transformation, developing vigorously the green economy and making positive progress.

Point three: Push green standardization in preserving ecological safety and developing green economy.

Along with the historic transition from traditional industrial civilization to eco-conservation, the standard system supporting economic and social development is also involved in a transformation to a green standard system required by eco-conservation. The so-called green standardization refers to the process of establishing and implementing the standards in accordance with the green philosophy of "ecological person" and the requirements of green economy. It is the crucial link that puts into effect the green philosophy, strategy and policy, and it is an extremely important link making the green economy practicable. For the sake of realizing ecological safety, we must accelerate the establishment and implementation of feasible, systematic, scientific and wholesome green standardization that covers every economic and social field and every aspect of life and production, and govern the society seamlessly with the green standardization. It is enough for achieving the best order and the greatest social and ecological benefits only by practicing and implementing the green standardization no matter whether people understand it or not.

It is also the successful experience at the international level in promoting ecological safety by using the green standardization. In an effort to deal with the more and more serious global challenges from energies and environment, China has established and implemented step by step numbers of laws, regulations and standards embodying principles of green economy, eco-conservation and environmental protection. As a region where there are many highly energy-consuming and highly-polluting industries, Shanxi province in recent years has actively established and implemented the green standardization, accelerated the transformation and upgrade of traditional industries, promoted the recycling economy, cleaner production and application of low carbon technology, and advanced the environmental protection and ecological restoration. Taiyuan, capital city of Shanxi Province, has now established and implemented 45 local green standards and is the city boasting of the most local green standards in China, thereby advancing effectively the growth of green life style and mode of production.

A green revolution bursting forth from the depth of soul is needed in preserving ecological safety. Let us make joint efforts to accelerate the promotion and realization of the revolution!

Thank you!

Climate Change and Ecological Safety [*]

Rusak O. N., President of International Academy of

Ecology and Life Protection Sciences (IAELPS)

I. Climate

The word "climate" was first employed over 2,000 years ago by Hipparchus, the astronomer of ancient Greece. The Greek counterpart of "climate" is "inclination".

Scholars defined this concept as "climate being the inclination of the earth's surface to the sun", which was believed to account for different kinds of weather of different longitude.

Climate is the statistics of weather phenomena in a certain region. It is determined by the geographical position of this region.

Global climate, which represents the earth's overall climate during certain years, is different from the regional climate.

Chart 1 presents the earth's climate system:

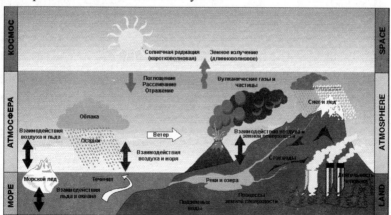

Chart 1 Components, forming process and interaction of basic climates

*   This article comes from the speaker's PPT.

Formation of climate includes factors of solar ray, atmospheric pressure, heat capacity, wind direction and power, air temperature, ocean, epigeosphere, etc.

And climate change is a fact confirmed by science.

Chart 2 presents the earth's climate change during a period of 22,000 years.

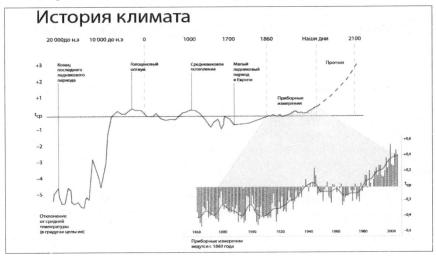

Chart 2 The earth's climate change during a period of 22,000 years

Chart 3 presents the average temperature of the northern hemisphere during the recent 5,500 years. We can see that the cold period is longer than the warm period.

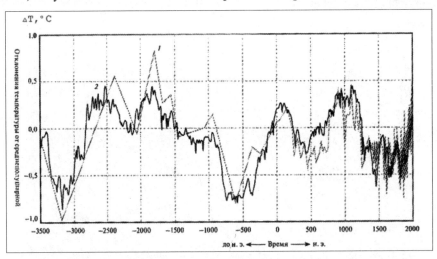

Chart 3 Average temperature of the northern hemisphere during the recent 5,500 years

Climatology is a science that studies the climates in history by using methods of dendroclimatology, SPAD value, glacioclimatology and historical documents etc. It is a must to know the past climate phenomena in order to understand the climate change of modern time.

II. Consequences of climate change

Natural disasters appeared with the climate change. 15,000 to 20,000 years ago, the humanity experienced the coldest period of all time.

It was during this period that humanity made great progress. They began to settle down on almost all the continents, learned to use fire and developed languages.

Though the climate is severe then, it was a period of great success and the prelude to a tranquil and subtle golden age.

The change of climate also led to the vicissitudes of civilizations and cultures. This is the idea of the geographic position determinism, which was proposed by Jean Baudin in 16th century, and supplemented and developed by L.N. Gumilev.

On the Eurasia Continent, climate change brought about drought and flood which followed each other in rotation. Climate deterioration in history not only resulted in migrations of ethnic groups, fall of dynasties, but the cause of poor harvest, famines, diseases and epidemics.

Chart 4 presents consequent scene graphs of climate change.

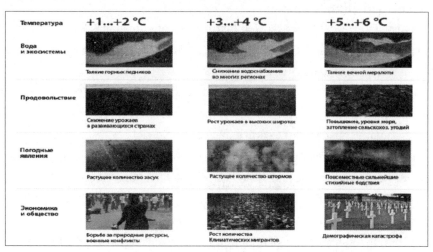

Chart 4 Possible consequent scene graphs under the circumstances of global warming (from text of speech of Stern, published on Summary of Economics of Climate Change in 2006)

Climate changes of modern time will be presented in Chart 5.

Chart 5 *Signals for global warming of modern time (map of climate change, Kristin Dow and Thomas E. Dowing, 2008)*

III. Causes of climate change

Statistics of recent years showed that the global climate has been changing and produced negative consequences. Climatologists' discussion on the causes of climate change is going on and there exist two theories:

Scholars of the first group insist that climate change is closely connected with events happened in the universe and on the earth, such as the solar and volcanic activities, polarity change of the earth's magnetic field etc.

However, a great many of scholars firmly believe that the humanity activities are the source of climate change and 90%-95% scholars hold this view.

The basic cause of climate change is the "greenhouse effect", resulted from the increase of carbon dioxide, vapor and ammonia in the atmosphere.

Indeed, the methane concentration in the atmosphere has reached to 150% and the carbon dioxide has increased by 30% since the industrial age. Though the green house gases are few in the atmosphere and occupy only about 1%, they play an important role in the formation of climate.

Chart 6 The formation process of greenhouse effect on the earth

Temperature on the earth will drop 30 centigrade without the greenhouse gases.

According to World Meteorological Organization (WMO), the average temperature of the earth's surface before the industrial age is 13.7 centigrade. The temperature should be 19 centigrade without the greenhouse effect, but now the surface temperature has reached 14.5 centigrade.

However, we believe the existence of the third theory, which is that the current climate change is the production of the coaction of nature and humanity's activity, only that the latter' influence is more obvious.

IV. Global warming and the consequences

According to the result observed with instruments, the warmest period is from the year 1995 to the year 2006.

See Chart 7 to know the change of global parameter variation.

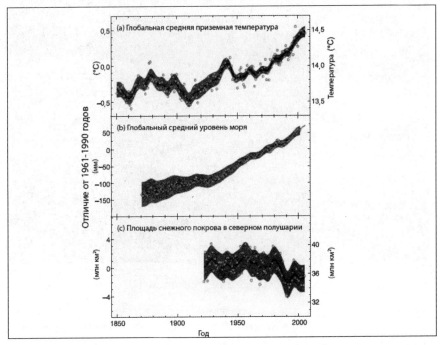

Chart 7　The change of temperature, sea level and areas covered with snow

Consequences resulted from the change of temperature, sea level and areas covered with snow:

  — increase of the probability of drought;

  — the instability of the soil and landslide;

  — meltage of the mountain glacier;

  — extinction of some ecological systems;

  — extinction of some species of flora and fauna;

  — quality degradation of water and food;

  — the appearance of climate refugees;

  — the appearance of some diseases;

V. The imminent global cold

The sun is the most important source of energy, and the quantity of energy which enter into the earth is decided by the change of sun range. If the sun range becomes shorter, global cold will arrive after 15 years to 20 years; conversely, global warming will arrive again.

The sun' area reached a summit at the end of 1980 and began to decrease in 1990, this means that the temperature will drop.

But we haven't sense this change yet, for the earth has absorbed too much heat. It is predicted that the global will become colder in 45-60 years from about the year 2050-2060 (+/- 11) , and then become warm again.

Diseases that global warming may bring about include bird flu, Babesiosis, cholera, Ebola malaria, Lyme disease, pestis, red tide, sleeping sickness, pulmonary tuberculosis and yellow fever.

VI. Role of humanity activity in global climate change

V. M. Kotlyakov, academician of Russian Academy of Sciences, holds that, among the factors that have an influence on climate change, the sun's influence occupies 83% and human activities only occupy 2.5%.

Most scholars believe that human activities have little influence on the climate change, and many of them support the theory that it is the space objects that affect the climate on the earth.

Some scholars think that, at present, there are not enough data to make exact prediction, and we must keep on observing, stick to objective study and refuse one-sided speculation.

The development of science must orient to the reason explanation and corresponding measures. But it is difficult for science to influence the space objects though it can influence and restrict the human activities.

VII. How to reduce the influence of human activity

1. Find ecologically safe energies.
2. Dispose carbon dioxide produced by the combustion of flue gas.
3. Increase the utilization ratio of renewable energies, now the ratio is 4%.
4. Apply energies from tides and waves.
5. Apply wind power.
6. Apply solar energy.
7. Apply energies of terrestrial heat and hydrothermal.
8. Develop nuclear power.
9. Apply controllable nuclear thermosynthesis. Russian has been doing the study

on this aspect for 25 years, but now it not carries out smoothly.

10. Develop biological energies, for example, wood waste, and manufacture bio-fuel, for example, bio-pellet.

VIII. Unique concept about alleviating threat of climate warming caused by natural factors

1. Artificial volcano.

2. Atomization of aerosol sulphur in the atmosphere.

3. Speculum that surrounds the earth (Million, d=60cm) .

4. Disk-like spatial installation that can reflect sun rays.

5. Nurture sea plants that can absorb carbon dioxide, and spray materials that help the plants grow in the sea.

6. Artificial ice-making in polar region.

7. Reduce the use of paper in an effort to control deforestation. This was proposed by Al Gore.

8. Other measures.

IX. Human factors in the international relations under the influence of climate change

Climate change has to do with every nation and country in the world. Established by the World Meteorological Organization (WMO) and United Nations Environment Program (UNEP) in 1988, the Intergovernmental Pane on Climate Change (IPCC) composes of thousands of scholars of various countries in the world.

And the United Nations has developed a great deal of work about climate change. In 1972, the UN General Assembly held the Conference on Environment and Development in Rio de Janeiro, at which heads of 156 states signed the United Nations Framework Convention on Climate which took effect on March 21, 1994. Members of the Convention has now reached 190 countries including Russia. It focuses on the reduction of greenhouse gases and avoiding the danger of climate change with concerted efforts.

Experts think that the discharge standard of 1990 is safe, and this standard must be safeguarded. (Chart 8).

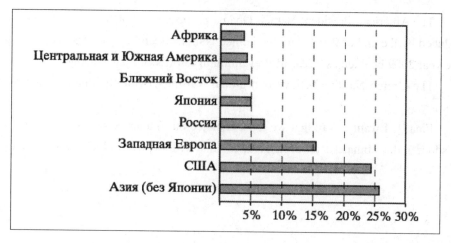

Chart 8 Global emission map of carbon dioxide in 1990

The United Nations Framework Convention on Climate was the first world-wide attempt to restrict the emission of greenhouse gases. But soon the reality showed that the Convention alone cannot reduce the emission, thus in 1997 the Kyoto Protocol, which legally established the goal of emission-reduction for the developed countries before 2012, was added to the Convention. And then, a new type of commodity — Emission Allowance for Greenhouse Gases — appeared in the international market.

Russia signed on the Protocol in 1999 and validated it in 2004.

The United States signed on the Protocol in 2001 but refused its validation. With a population occupying 6% of that of the world, the United States occupies a 25% of carbon dioxide emissions across the globe.

Even to this day, the countries' attitudes towards Kyoto Protocol are still vague.

And it is widely believed that, instead of aiming at the climate, the Kyoto Protocol is only a political agreement.

In 2007, the 13th annual climate conference held at Bali approved a new international agreement to replace the Kyoto Protocol which would be due in 2012.

The 15th Copenhagen Climate Conference held in December of 2009 concluded in discord and without any fruit.

In 2007, British submitted a special bill on climate change to the United Nations Security Council. Nothing emerged though a public hearing was held. On December 17, 2009, Russia approved a climate report in which a national policy focusing on global and regional climate change and its consequences was fixed.

The American President Barack Obama proposed that 80% of the emissions be reduced by the end of 2050. And the United Nations Secretary-General called on that the year 2009 be fixed as "Year of Prevention of Earth's Climate Change".

The United Nations believe that global warming is the dominant theme of 21 century.

Finally, I want to say that we must enhance our humanity's thinking ability and wisdom in this climate-deteriorated era, and this will help us walk on a correct path.

Pushing Forward Eco-Safety Requires Two Keys

Yang Shen, Chairman of Experts Committee of IESCO and

Former Vice Minister of the Ministry of Construction

Chairman of the Assembly,
All the Distinguished Guests,
Ladies and Gentlemen,

In accordance with the arrangement of the Assembly, I'm going to make a speech on city eco-safety and disaster early warning to elaborate my several points of view with the help of Shanghai World Expo stage, which just closed.

Shanghai World Expo, lasting for half a year, closed on October 31. This World Expo on the theme of city held for 184 days with 246 participating countries and international organizations and with 73 million visitors and 390 thousand visitors on average a day. This is the first expo held in a developing country since 159 years ago.

On the closing day of World Expo, Shanghai Declaration is broadcasted to the world, which is formed by the sponsor units, World Expo Bureau officials and participating units through concerted negotiation. It seriously declared that October 31 is the annual "City Day", called on people to rethink the relationship among people, city and the earth from the perspective of new historic height. The Declaration points out that what urbanization and industrialization have brought for human beings is not only modern civilization achievements but also unprecedented challenges at the same time, for example, population expansion, traffic jam, environmental pollution, resources shortage, city poverty, cultural conflict and so on. By reading between the lines, we can get the predicament faced by developing countries and their expectation of future.

The biggest highlight of Shanghai Declaration is that the concept of maintaining eco-conservation and promoting eco-safety elaborated by it is almost same or identical with the objectives of World Eco-Safety Assembly. The Declaration put "the creation of eco-conservation with the needs of future in mind" in the first place, fully expressed the common aspiration of developing countries and all countries in the world to strengthen eco-safety.

Ladies and gentlemen, as a staff being devoted to eco-safety, I think maintaining eco-safety is the holy responsibility of all mankind and strengthening disaster early warning is our common obligation. I recommend that these things could be pushed forward from the following two key points:

First, strengthening theoretical system building. Passing through long history, from prehistoric civilization mankind evolved into agricultural civilization and industrial civilization and now enters into the scientific and technological civilization era. However, scientific and technological development is endless. It can be predicted that mankind will usher in the ecological conservation era in the near future. The fundamental characteristic of ecological conservation is that social productivity is highly developed and mankind and nature could develop harmoniously. This is also the goal that this World Eco-Safety Assembly is striving to realize.

For that reason, we must highly consider the primary importance of the popularization of eco-safety concept and extensively popularize eco-safety knowledge to the civil servants at all levels first, then to schools at all levels and public occasions to promote ecological building. At the same time I suggest getting a newspaper International Eco-safety Guide afloat, setting up international eco-safety website, popularizing eco-safety knowledge through various media, and promoting eco-safety knowledge level of employed persons and the public.

Second, insisting on pushing forward according to the law. Eco-safety is a public welfare undertaking and can benefit the people and the country as a whole, so we must mobilize and organize the public to participate in this undertaking actively. When we popularize ecological knowledge, we should form a whole set of law and regulation system to promote the eco-safety course by law. Thereby, we must strengthen legislation, enforce the law strictly and recommend International Court of Justice in Hague to establish international ecological court to standardize people's behavior by laws. In the society and world we could bring about a systematic environment in which everyone knows the law, obeys the law and maintains eco-safety by laws.

In a word, education and law are the two magic keys to push forward ecological safety. The above is my speech. Wish the Assembly a great success. Thank you!

III
Thematic Forum
(Ecological Safety and Disaster Early Warning)

Find the Best Development Pattern in the Scarcity of Resources *

Rio D. Praaning Prawira Adiningrat, Secretary-General of PA

(Europe) International Foundation

Our world has around 6.8 billion inhabitants. Of those, approximately 925 million live through a day without adequate food. The reasons vary. They can emanate from natural or manmade crises. These crises can be systemic or incidental. But they all have in common that there is scarcity.

Scarcity is an economic phenomenon. Experts state that economics are the science of scarcity. Indeed if we discuss economic models of today or in the past the successful or failing management of scarcity is the ultimate measure of economic policy.

In today's globalized world economic policy and thus scarcity are clearly and directly influenced by both domestic and international developments. Perhaps in the past it was not as clear as today, but the global economy is clearly involved in a massive rebalancing act. This implies that funding and investment are more than ever drawn to places where the return on investment is best.

Foreign Direct Investment, Return On Investment and the management of scarcity create new pressures and resulting security concerns in one and the same global region. It is useful to understand how such coincidences have occurred and evolved in the past and how lessons can be drawn for the future.

Quite apart from political/strategic/ideological angles, elements of preparedness and prevention can mitigate any tension; but if these take the shape of increased levels of defense spending and/or military presence, tensions may go up.

At the same time, official development aid from the North to the South is

---

* This article comes from the speaker's PPT.

declining and is expected to be reduced further and stronger. Aid recipients are increasingly considered trading partners and why would you assist your opponent with a better deal? Yet the scarcity factor remains a decisive indicator for the true humanitarian issues at stake. How to align these two different factors?

An equally uneasy question to be addressed is the relationship between formal foreign and defense policy, industrial policy and the relationship between Government institutions and industries. At the same time, Sovereign Funds play an increasingly important role in national economies globally. Last but not least, the sheer volume of global flows of Trade and Investment suggest that there should be a stronger basis for Corporate Social Responsibility and the development of economies and markets in well understood mutual interest of both civic society and industry.

Many claim that CSR should be part of a new type of Public Private Partnerships. If this is correct, it could be equally correct to claim that a new link should be made between official development aid and Corporate Social Responsibility.

These days we are once again confronted with the reality that exchange rates and 'quantitative easing' influence trade, employment, income and thus taxes, healthcare, education, defense spending, environmental protection etc. Monetary policies and protectionism also impact on feelings of inse curity and thus of scarcity.

So how do these financial, economic, development aid, defense spending etc. phenomena hang together? Below follows an overview of statistical material of the World Bank, the IISS, WHO, UNDP, etc. identifying how different flows and numbers hang together as time passes. Of course there are lies, damn lies and statistics, so flow models below are only indicative and try to make comparisons possible.

The above does suggest that there are intensive changes in the balance between the different parts of the world. In fact there may be three interacting and changing balances: the global military balance, the global trade balance and the global ecological balance. These three global balances interact and in so doing produce an environment for each individual civilian at whatever place on earth through an individual perception of personal security. While these global balances may have their influence on the individual civilian, this person perceives his personal security situation on the basis of his personal ability to deal with issues of scarcity: can the person provide sufficient shelter, food, health and education for his or her family.

By far the majority of individual civilians is employed through small or medium-sized firms. The income derived must provide for the livelihood of the family. Hence the enormous importance of connecting SMEs through elements of trade, investment, Corporate Social Responsibility, Public Private Partnerships, etc. to policies that seek to promote the viability and success of the SMEs.

If the above is true in times of relative peace, stability and lack of human or manmade disasters, it is even more true when calamity strikes. Again, natural disasters cannot be prevented but civic society can be prepared. The same applies to post-conflict civic society management. In fact, the level of preparedness and prevention of such disasters is decisive for the capacity of the individual citizen to cope with the consequences.

We carried out the prevention and implementation around two years ago and prevented a catastrophic ban of all fish export from an Asian country to the European Union. The mixture of Public Private Partnerships, Corporate Social Responsibility approaches and official development aid not only prevented a fish export ban but created the environment for a substantial expansion of this fish export to the European Union – thus contributing to employment, improved per capita income, etc.

In conclusion the following can be stated:

a. The global rebalancing act in financial, economic, military and ecological terms includes a stinging scarcity element that requires in-depth analysis potentially leading to improved international solutions that may also avoid protectionism.

b. A holistic view at the interaction between Governments and industries in the global economy leads to the requirement to study ways of integrating official development aid with Corporate Social Responsibility approaches potentially through specific Public Private Partnership programs.

c. The interaction between the globally changing military, economic and ecological balances and the resulting concept of personal security require a re-focusing at the importance of the promotion of the establishment of viable small and medium-sized enterprises.

d. Natural disasters cannot be prevented but civic society can be optimally prepared. The concept of personal security finds expression through the prepared promotion of the establishment or recovery of small and medium-sized enterprises in post-disaster or post-crisis areas. Preparedness and prevention can be explicit elements of an integrated approach of national and international development aid, emergency assistance and Corporate Social Responsibility.

Find a Best Solution for Harmonious Development of Mankind and Nature

Vesselin Popovski, Senior Academic Program officer of United Nations University

Let's first of all, give thanks, especial my institutional and personal thanks to Professor Dr. Jiang Mingjun for making this happen today and yesterday. The Assembly, I think, is a historic step and it is the first World Assembly of Eco-Safety which brings us to the core of the vital issues which we are discussing nowadays.

I would like to tell you what I think about, why first natural disasters are no longer natural disasters, they are not only natural disasters, and I would call them unnatural disasters, because human beings already can affect the nature to the effect that they can actually destroy the nature. We are living in the moment when we are all setting the relationship between nature and human. Historically, all the human progress was about protecting human from the nature, from natural floods, earthquake, and disasters. It's time to protect the nature from the humans and the power of the human beings can destroy but also can save the world.

We have excellent presentation by Professor Rusak on the history of the development of the climate debate, and how nature could balance the head of our life. Now the question will be whether humans can also balance the life. And that's why, I think, the first World Eco-Safety Assembly is a tremendous historical moment. We need to ask this question. Can humans protect the nature, if nature over millions of years protected the humans? What's the balancing factor of life? Here, it comes to the responsibility for eco-safety, for protection, from unnatural disasters. I very much enjoy the four conclusions which Director-General Jiang Mingjun just presented well, because in our research we also found that it needs a holistic approach for common action.

We need public sectors, which comprises government, civil society; also the private sectors, which comprises business corporation; but not only, here come the individual people and individual entrepreneurs and consumers, can also play a major role. So, we have basically four types of actors and four types of responsibility for protecting eco-safety. Governments, civil society, businesses and individual humans, public sector, politicians, civil society and private sector, businesses and individual entrepreneurs, somehow in the middle of those four comes the science. What sciences can pay in order to keep the balance between nature and humans? What I found particularly worrying and problematic is Dr. Policy-Maker; he will keep the same discourse. They are somehow natural to win elections. They need to talk to economic growth first, maximization of power, maximization of wealth, doing more, producing more, and consuming more. That maximizing language is absolutely the opposite of what eco-safety requires.

We need to talk about the opposite of consuming less, producing less, less energy, less carbon and even less children. I hope politicians will change their discourse from more and more to less and less. Somehow the human happiness has always been associated with big cars, spending a lot of fuels, big families, big estates and restaurants, big cities for facility. But this is exactly again is the opposite for the future demands.

The future demands little family, little car, less meet and consumption, less facility on the streets. I think one of the answers might be found in scientific community. We have such great scholarship here from china, from Russia, from Southeast Asia and South Asia. Scientists have already provided some of the answers. Somehow we should ask people to reduce their lifestyle. We could offer them green technology, less carbon, low energy way of life, which still can satisfy human happiness.

Here, what I think is important is also to have the synergy between ICAPP political parties in the same building with the World Eco-Safety Assembly can really inform the policy maker. Particularly Asian way of life has been always very harmonic. It was American and European around the model of happiness, based on capitalism, based on over consumption, consumerism that produces the problems we are facing today. Here comes an opposite Asian model of happiness which has less to do with capitalist, growth of maximization of power, wealth and over consumerism, happiness with the nature, living in harmony with nature. And I see that moment of Asia scientific community influencing their own policy-makers first and then globalizing the fact of they are finding of how to live in harmony with nature. As one of the solutions, how future policy-makers will win elections by telling their electorates not to produce more, consume more, have more children, have big car and big estate, but the opposite.

Thank you very much!

Address Urban Disasters and Conflicts with "Ecological Wisdom"

Dong Jinyi, Deputy Board Director-General of China Foundation for International Studies

All the respected guests,
ladies and gentlemen,

Hello! I am very glad, upon invitation, to attend this World Eco-Safety Assembly and give a keynote speech. Centering around the theme of "climate change and eco-safety", I'd like to talk about how to address urban disasters and conflicts with "ecological wisdom."

Ecological safety is an important force that carries forward sustainable urbanization and a new subject in international cooperation and exchanges over the recent years. Confronted with the various kinds of disasters and conflicts during urbanization, we need to guide urban sustainable development with new eco-safety concepts.

The natural variation on earth (including natural variation triggered by human activities) occurs anytime and anywhere. When it jeopardizes the human society, it becomes disaster and conflict. Disaster and conflict is a kind of manifestation that human and nature contradict with each other and has double attributes of nature and society. It is one of the most severe challenges faced by human beings in the past, at this moment and in the future. Significant worldwide disasters and conflicts mainly include droughts, floods, hurricanes, storm surge, freezes, hails, tsunami, earthquakes, volcanoes, landslides, mud-slide, forest fires, plant diseases and pests etc.

Significant natural disasters in a country used to be viewed as local-bound with limited effects, and prevention and reduction of these disasters were expected to be the responsibility of the country alone. Now, economic and information globalization have enabled countries to be more mutually dependent and more closely linked, and major natural disasters in a country or region are likely to involve others in trouble and make negative effects on national stability, development and safety. For example, Enormous economic loss and expenditure and inner fear of the public and social panic. Ineffective emergency control and post-disaster relief may trigger off public riot, social unrest and even political crises which may affect national stability. Besides, major domestic disasters would force a country to concentrate its attention on the efforts to cope with the domestic issues, thus reducing its capability to cope with major external events and crises that are related to its national safety.

Just since this July, because of attack by the torrid summer in Russia, a spate of forest fire disasters of large scale are triggered off with over 550 forest burning points, more than 20 mudstone burning points and the total burning area of 190 thousand hectares. Lasting smoke and fog caused air pollutant volume in Russia to reach a new high, and for each day the death toll is up to about 700, which has increased onefold compared with the corresponding period of last year and caused panic among the populace. Since late July, because of continuous rain, Pakistan has suffered from the most severe flood disaster in the past 80 years. As is estimated, flood ruined a huge amount of civilian residences, roads, bridges and farmlands. Because of the flood, at least 1600 persons died; 15 million persons suffered from disasters, and over 4 million persons become homeless. In the early morning of August 8, 2010, suddenly heavy rain fell down on Zhouqu County, Gansu Southern Tibet Autonomous State, China. The mud-rock flow of Luojia Valley and Sanyan Valley in the north of county surged down and rushed into the county, so the residences along the river were destroyed. In this mud-rock flow disaster over 1200 persons died; 588 persons disappeared; over 20 thousand persons suffered from the disaster. The casualty toll is equal to the sum of dead and missing people in various geological disasters, such as, landslide, surface collapse and mud-rock flow, in the past three years throughout the whole country.

Natural disasters can not be totally avoided, and despite our persistent fight against them by all possible means, they still happen. This does not mean, however, that we can do nothing about them. We need to take the initiative to address urban disasters and conflicts with "ecological wisdom."

The thoughts on "ecological wisdom" originated in China. Such intellectual fruits as "yin and yang", "man & nature harmony" and "Tao following nature" attribute themselves to efforts of the ancients in striving to live in harmony with nature. At present, disasters are frequent occurrence, bringing man persistent damage. While paying attention to disasters and discussing how to prevent and reduce them, man needs to review this old but advanced "ecological wisdom" in order to explore the life process of cities and their interactive relationship with external environmental systems, to distribute properly, to develop scientifically and to seek harmony among our life, production and nature. To better address disasters and conflicts, avoid them as much as possible and reduce their damage, we need to do the following jobs:

1. To sensitize disaster-prevention awareness. National governments should strengthen science popularization and educate the public on disaster prevention and reduction in every possible way to raise their safety awareness.

2. To perfect the emergency management mechanism. Disasters and conflicts often show some indications which, if obtained through scientific monitoring, can enable us to take preventive measures and avoid the disaster. But when they happen, we need to take measures as soon as possible to prevent them from growing and to reduce their damage as much as possible. That is to say, rapid actions are needed once a disaster or conflict happens.

3. To better prepare contingency plan against disaster and conflict. Such a contingency plan is an important part of emergency management. Once an emergency comes up, due measures can be taken in light of the contingency plan and of the actual situation.

4. To well manage contingency work against disaster and conflict. Once disasters and conflicts happen, a large contingent of personnel, a large amount of relief materials and capital are needed. For that reason, it is necessary to make professional personnel well trained and prepared, relief materials and capitals instantly available even in time of peace.

In a word, urban disasters and conflicts are unexpected, disorderly and harmful. It needs the efforts of governmental departments at different levels and the various social forces and even international cooperation to jointly prevent, lessen and cope with a disaster through establishing scientific systems and preventing it from its source.

Ecological Safety Problems When Fighting Land Desertification

Ugodchikov G.A., President of International University of

Business-Technologies, Moscow, Russia

Dear Ladies and Gentlemen!

I am sincerely grateful to the organisators of the first worldwide UN assembly of ecological safety for giving me this opportunity to speak. I am a member of the Supervisory board of a Science-city of Russian Federation, namely the city of Michurinsk since the date of it's foundation on Nov. 4, 2003, and I am also the bearer of knowledge of Soviet and Russian scientific schools on the controlled biosynthesis (controlled cultivation of cells of microbes, plants and animals). One of the founders of this particular scientific school was the well-known academician I.N.Blochina (1921-1999).

Science-city of Michurinsk, Russian Federation, was created for the purpose of innovation technologies development in the agricultural domain and their further replication on the whole territory of Russia.

In the domain of controlled biosynthesis in the Science-city of Michurinsk, RF on the basis of State Educational University of Michurinsk, the following scientific production base was created:

— the controlled cultivation laboratory;

— experimental-industrial production facilities (pilot production);

— experimental fields of 10 hectares for the development of intensive cultivation of agricultural crops and organic waste utilization.

The specialized biotechnological equipment, namely the bioreactor-computers, was provided by the I.N. Blochina International Fund of Biotechnolgies, which is also directed by me.

Due to the availability of such biotechnological scientific-production base, it was possible to develop biotechnological solutions, specifically designed for the growth of vegetable and decorative crops on a sandy soil, which is particularly important in realization of the UN program to combat and prevent desertification, which is planned for 2010-2020.

It is also important to stress the fact that in implementation of these particular biotechnologies, there are no chemical regents such as fertilizers, plant protection means and etc. being used, which ensures that agriculture and farming remains truly organic. All technologies are patented and registered in Russia in 2010. The absence of chemical reagents in cultivation of plants guarantees high ecological safety.

It is also worth mentioning that the existence of the above-mentioned biotechnological scientific-production base in the science-city of Michurinsk, RF, allowed to develop ecologically safe industrial biotechnologies for the utilization of food waste and their further transformation into the highly effective organic fertilizers for growing plants, which could additionally be used the severe conditions of sandy soil or desert. In the city of Michurinsk, the food waste of fifty food industry enterprises such as cafes, restaurants, etc. were simultaneously utilized.

Fertilizers obtained as a result of such biotechnological processing may act as basis for creation of necessary bulk layer of soil above the sandy territories of deserts. This particular approach allows to replace such expensive and scarce raw material sources as peat and sapropel, which are usually used in the creation of the soil layer in the conditions of deserts. This approach was named "Waste and Revenue" which is perfectly true, as the food waste processing considerably improves ecological safety of cities and settlements, additionally decreasing the costs to combat desertification when creating artificial soil.

Another important positive quality of works effected in the science-city, RF Michurinsk is the development of working biotechnologies solutions designed to accelerate vegetable and decorative seed germination which is also extremely important in harsh environment of deserts and semi-deserts under conditions of accelerated erosion of soil elements.

Apparently, the creation of such scientific-production bases (centers) is appropriate not only in Russia but also in other countries, which are actively interested in implementation of UN program to combat and prevent desertification. The first step in this direction was the development and signing of a draft project "implementation program of industrial biotechnologies into the agriculture an environmental protection within the framework of UNIDO system". The project was signed on Aug. 14, 2009 by the authorized representatives of the authorities of the Science-city, RF – the head of administration, doctor of agricultural sciences V.N. Makarov on one side and the authorities UNIDO Moscow office – director S. A. Korotkov on the other side.

However, the issue of funding of these works by UNIDO is still not resolved, and as a result, the works are conducted at personal expense of the first party that signed the project.

In order to accelerate the implementation of innovative technologies, in structure of International University of Business-Technologies in November of 2009, the International Institute of Breakthrough Technologies and Innovations (MIPTI-IIBTI) was created, which performs its works at the intersection of such sciences as Physics, Chemistry, Biology and etc. for different industrial and economic sectors: new materials, mechanical engineering, transport, industrial biotechnologies in agriculture, medicine and environmental protection. The chairman of the board of directors of MIPTI – Professor S. E. Moysis. We invite any interested parties for cooperation.

As concluding remarks and on this high-level platform, I am about to make suggestions to all of you. As previously mentioned, under the United Nations framework of desertification control project in the next ten years, a great many of international or multi-national cooperative projects will be launched.

I propose that in our next agenda (after the thematic session) we discuss the issue of holding international desertification conference which is predicted to be held between the years 2012 and 2013. All countries that take an interest in this topic are expected to be present.

Thank all the present for your attention.

Build Lingyun Mountain into a Best Practice of Eco-Tourism in the World

Hu Wenlong, Secretary of the Commission for Discipline

Inspectionce of Nanchong City, Sichuan Province, China

Respected guests,
ladies and gentlemen,

We feel quite honored to be invited to participate in the First World Eco-Safety Assembly in beautiful Phnom Penh. With the increase of world population, the exploitation and use of resources, there is more and more serious damage to the environment caused by human activities, which poses an increasingly serious threat to ecological safety. This situation has attracted extensive attention of the governments of countries across the world and that of the international organizations. Though, in recent years, various countries in the world have taken effective measures in preventing natural disasters and dealing with eco-catastrophes and achieved remarkable results, the trend of environmental deterioration has not been reversed, and the environmental disasters and eco-catastrophes resulted from environmental degradation and ecological crises cannot be effectively checked. Therefore, all is facing the severe challenges of ecological environment regardless of individuals, regions, or countries. It has become a common consensus to preserve global and regional ecological safety. People understand more and more deeply that only ecological safety is preserved by strengthening cooperation and sharing resources, can the win-win situation be realized.

Nanchong City is located in the south-west China and the middle reaches of the Jialing River. It covers an area of 1,250,000 square kilometers, with a total population of 7,560,000. It always has a good reputation of "Long-standing Silk City" and is hailed as a shining peal on the Jialing River, which is the largest tributary of the Yangtze River and the "Mother River" of Nanchong. It flows through 7 cities, districts and counties and extends as long as 298 kilometers. Along the sides of the Jialing River, grass and trees grow luxuriantly, and birds are flying with full vitality. There live 2,000 kinds of wild plants and 665 kinds of wild animals. The eco-tourism area of Jialing River Valley is one of the "New Five Tourist Areas", which are the key construction projects during the 11th Five-Year Plan period of Sichuan province and of which the Nanchong Sector of the Jialing River is the core area. The Lingyun Mountain Scenic Spot is facing the Jialing River and is receiving the major emphasis in the development and construction of Jialing River Valley. It now has been successfully built into an AAAA National Scenic Attraction and an eco-tourism demonstration base. In the development and construction of Lingyun Mountain Scenic Spot, we have always stuck to the development law of harmony and co-existence between human and nature, and the sustainable development of keeping balance between conservation and development. The originally wild mountains in the suburbs were planned and constructed into an eco-tourism project featured by ecology, culture and religion, where the hills and the waters enhance each other's beauty, man and nature become interdependent. It embodies the concept of ecological civilization and safety, and seeks a trinity development mode of economy, ecology and society.

Keep balance between conservation and development, enrich the ecological resources of Lingyun Mountain. There is a forest coverage of 86% there. The mountains are rich in various plants, thousands of bushes and trees, and nearly one hundred kinds of wild and rare animals. Lingyun Mountain Scenic Spot also has two Buddhas (the sleeping one is 72 meters high and the standing one is 99 meters high), three mountains (Baishan Mountain, Tushan Mountain, Lingyun Mountain), four seas (forest sea, cloud sea, flower sea and bamboo sea) and five lakes (Lingyun Lake, Baishan Lake, Tushan Lake, Xuanwu Lake and Junior Xihu Lake). It is not only the National Forest Park and Provincial Geological Park, but also the Cultural Industry Demonstration Base of Sichuan Province and China's Top Scenic Spot of Geomantic Quality. The rich resources of primitive forest are the most important natural heritage of Lingyun Mountain Tourist Attraction, as well as the important foundation for building Lingyun Mountain into an International Eco-tourism Demonstration Base.

Inherit and carry forward the religious culture of Lingyun Mountain. The coexistence of Confucianism, Taoism and Buddhism get perfect annotation and concentrated embodiment in Lingyuan Mountain Scenic Spot. Until today, all the remaining temples, monasteries and pavilions in the Lingyun Mountain since the Han Dynasty and the Tang Dynasty have been kept in good shape. An important religious activity — "the pilgrimage to Soul Mountain on the third day of the third month on the Chinese lunar calendar" — recorded in the documents has been still in practice. The religious culture here has a very long history and has an important value for protection and study.

The mind and the nature interact to enjoy the geomantic culture. Centered on the main peak of Lingyun Mountain, the surrounding hills form the shapes of dragon, tiger, tortoise and rosefinch of old legend. They correspond to the arcane truth of "four animals and five elements" in the ancient Chinese art of geomantic omen and naturally take their positions according to the geomantic model (the front is the rosefinch, the back is the black tortoise, the left is the green dragon and the right is the white tiger). All of them imitate the living things to the life and become the unity of the form and spirit. They are considered the live specimen of the Chinese geomantic geography and have the great value of study.

In March of this year, IESCO confered the title "International Eco-tourism Demonstration Base" upon the Lingyun Mountain Scenic Spot. This affirms fully the practice of sticking to the concept of harmonious development of economy, society and ecology. I hope that IESCO may pay continuously the attention to the development and construction of Lingyun Mountain, give great support in the aspect of funds, technology and publicity in an effort to build the Lingyun Mountain Scenic Spot into the best practice in protecting natural heritage and developing harmoniously the ecology, economy and society.

Respected ladies and gentlemen, Nanchong is the center city of north-east region of Sichuan province, China. It has attracted the world's attention because of its ecological construction, which is also the bond of forming a connection with IESCO. We warmly and sincerely welcome the political VIPs and foreign friends to visit Nanchong, welcome the industrialists to invest in and establish businesses in Nanchong in order to promot the cooperation on trade and ecology and to create a beautiful future!

Finally, wish the Assembly a great success!

Thank you!

International Ecological Safety and Disaster Early Warning

Chak-Kwong Au, Trustee of the Board of Education of

Richmond City, Canada

Due to the worsening eco-safety environment, international cooperation is more important than any period of our time. Ecological safety issues affect not only a particular country or region, but every county on this planet. The disasters caused by ecological damage are usually wide-spreading and has a long lasting effect. The Gulf of Mexico Oil spill in April 2010, for example, threatens not just to the US and nearby Ocean. Official estimated the leak amount of oil totally around 114,000 barrels (18,100 m^3). This 3 months oil spill has caused extensively damage to marine and wildlife habitats. It is estimated that there is 2 inches thick layers of oily material covering the bottom of the seafloor, more than 30% of fisheries were killed, and it has threatened more than 400 species. Experts are still evaluating the ecological damage caused by this oil spill, but marine ecology, for sure, has been affected by this spill.

Although not all ecological crises can be as worse as the Gulf of Mexico Oil spill, all can cause the cross-broader crisis. The benzene was spilled into Songhua River because of an explosion at a petrochemical plant run by the China National Petroleum Corp's Jilin Petrochemical Company. The spill caused a great concern for Russia; Russia's afraid the polluted water would reach Amur River in Russia. Emergency early warning systems, in this case, were crucial to handle the crisis. A minor environmental accident can have a cross boarder effect if those toxic and harmful substances release to river stream, underground water and atmosphere. When the ecological crisis break out in a country that situate in upper reach of the river, it inevitable affects the country that is located in lower course of the river. When the accident leaves unnoticed; the monitoring system in lower reach country is ineffective; or the toxic and harmful substances cannot be detected, people living in lower reach of the river may become victim of ecological accident; or it may even be the source of conflict.

Some minor accidents, at first, may not have immediate impact. However, if we treat it lightly, the consequence may be serious. For example, some toxic substances can leach to underground and spread around through underground water. We may not notice at first, when the toxic substances accumulate to certain level, we may have already missed the best time to impose essential mechanism to stop the crisis, and caused the economic damage and the loss of human lives.

Even though the ecological crisis can no longer be ignored, the cooperation among governments is still facing a lot of barriers. For example, we haven't reached a binding agreement on cutting carbon emission or built an early warning system for eco-disaster. As lacking of early warning system, BP was accused of delaying information and hiding crucial information after the oil spill in Mexico Gulf, and it caused the US and International community unable to impose necessary measures to stop the spill immediately. In order to stop the environment worsening, we need to cooperate and to eliminate the roots of environment deterioration. Also an effective early warning system is also the key since it can stop the spreading, and minimize the damage as early as possible.

There is not lacking examples of international community building mechanism in order to ensure mutual interests. IAEA, for example, established "The Convention on Early Warning of a Nuclear Accident"; "Convention on Assistance in the Case of a Nuclear Accident or Radiological Emergency", "Convention on the Safety of Spent Fuel Management and on the Safety of Radioactive Waste Management", "Convention on Physical Protection of Nuclear Material" and so on. Those measures strengthen the safety use of nuclear energy, international cooperation, and early warning system. Also, WHO also established collaborating centers in different regions to collect, collate and disseminate information, which has proven that those centers are capable of monitoring and preventing the widespread of epidemic. I think International Eco-safety Cooperative Organization can learn from those excellent examples.

All countries should understand when the ecological crisis breaks out; no country could stay out of it. Establishing early warning system for ecological crisis are necessary and unavoidable. Mutual trust and mutual assistance are the foundation of those establishments. At least, 3 main aspects need to be included into International cooperation:

1. Responsible Attitude: Information about ecological incident should not be "Classified" and should inform the neighborhood countries and the globe without delay, in order to reduce the risk of spreading the crisis.

2. Mutual Assistance： If needed, country should accept help from international community, and handle the crisis together, in order to minimize the damage.

3. Information Sharing: All information related to crisis should be organized and shared after the crisis, in order to strengthen the collective capability for crisis management operation.

Mutual trust can be built only through communication and dialogue. Thus, it is necessary to build a high-end platform, like this Assembly, for communication and further cooperation. However, it is not easy to reach a common ground on how to handle the ecological crisis, due to the differences in economic development, political structure, geographical location, and culture. It will take time and patience to build mutual trust. To begin with, I would suggest building a regional mechanism since they have more in common. We can then further the cooperation from regional to global level.

To build early warning system, some technical problems need to be solved. The most important of all is to standardize the evaluation because of the diverse and complex nature of ecological safety. We also should classify accident by its nature, type, fatalness, risk of spreading, and so fourth. We can only speed up the evaluation process, and make the alert system work by standardizing the evaluation procedure. If not, it would only worsen the situation.

Building a global ecological risk analyzing system, of course, is a challenging job. The only realistic chance of success is by cooperation among different research institutes, professional groups and academic institutions around the globe.

At last, every country needs to work on perfection of their monitoring system. If there is a loophole on the local monitoring system, an effective international cooperative system on monitoring of ecological crisis is just a romantic notion. So the developed countries should provide assistances to those countries which has no resource to develop a proper early warning system.

Ecological safety is already on the edge and can no longer be ignored. If international cooperation is still on the stage of paying lip service, we will miss the last opportunity to tackle ecological crisis. We can only turn it around by long-term cooperation in all fields, and establish early warning system. I hope we can see the blue sky, green land, and river packed with marine life soon.

Thank you!

Practice of and Reflecion on Tangxia's Ecology Construction

Ye Jinhe, Secretary of the Tangxia Town Party Committee, Dongguan City, China

Distinguished guests and dear friends,

Good Morning!

With the beautiful winter scenery of Phnom Penh, the First World Eco-Safety Assembly is held solemnly here. I would like, on behalf of the Party Committee and the Government of Tangxia Town, Dongguan City, China, to express my warmest congratulations on the convening of the Assembly. And I would also like to take this opportunity to express my sincere gratitude to the organizers — Royal Government of Cambodia, International Conference of Asian Political Party and International Eco-Safety Cooperative Organization — for their support and help given to us in our participation in this Assembly.

Located in the Pearl River Delta of South China and among the economic corridor of Dongguan, Shenzhen and Hongkong, Tangxia Town is the center town of Guangdong Province and the regional center of Southeast Dongguan. It covers an area of 128 square kilometers with a population of over 400,000. Bathing in the spring breeze of China's reform and open policy, the economic and social development of Tangxia Town has gained rapid growth. It has developed from an ordinary rural town to an important international processing and manufacturing town and World Golf Town, and has ranked the fifth in the one thousand towns with powerful economic development in China. In 2009, its GDP reached 16.87 billion Yuan, and this year's GDP is estimated to increase 10% on last year's basis.

Making a general survey of the industrialization and urbanization course of Tangxia Town, we can see that it always sticks to the scientific concept of development which lays equal emphasis on economy and environment and promotes harmonious coexistence between mankind and nature. While pushing forward the economic development with all its strength, Tangxi Town attaches great importance to the protection of natural ecology and development of social and public services, makes great efforts to carry out "green GDP project, blue sky project, clean water project, friendly living environment project and green land project", optimizes the layout of garden planning and enhances the allocation of gardens and green land, develops green low carbon economy and implements the policy of energy conservation, emission reduction and efficiency increase to improve its environmental quality, enhances its comprehensive sustaining capacity and strengthens its comprehensive treatment of social order and security to increase people's happiness index. In this way, Tangxia Town constructs a dynamic eco-safety system with surrounding ecological green land, circulation of green economy and harmonious civilization, creates a garden city of Lingnan flavor featuring "city among trees and people among green land", promotes the harmonious and sustainable development of economy, society and natural environment, builds an excellent environment which is fit for living as well as commerce, and realizes the town's planning and development objective that there are gold and silver mountains as well as green mountains and clear water.

So far, the town's greening coverage of the built area has reached 2318.02 hectares, occupying over 20% of the town's area, and per capita green area amounts to 15.6 square meters. The Green Corridor— the evergreen Yingbin Boulevard which crosses the central town, the natural protective screen—the Forest Park which stands in the southwest of the town and constitutes a natural urban oxygen bar, Tangxia's primitive ecological garden and landscape where Dazhongling Wetland Park is located, and Guanlan Lake golf club which is considered to be the largest in the world by Guiness all reflect the great harmony and unity between the living environment and natural environment.

Tangxia Town has made gratifying achievements in its eco-safety construction, and has been awarded the following titles: China National Garden Town, China Famous Green Town, China Beautiful Town and China Hygiene Town, and so on. On July 30 of this year, after passing the expert assessment organized by UN/International Academy of Ecology and Life Protection Sciences and International Eco-Safety

Cooperative Organization, it was granted the honorable title of "International Eco-Safety Demonstration Town", which marks the new stage of its eco-safety construction.

In view of the practice of Tangxia's construction of eco-safety town and sustainable development, we hold that the construction of eco-safety town is a comprehensive project, a civilized, healthy, harmonious and complex system full of vitality, a regional pattern with virtuous ecological circle and a realm where human and nature coexist harmoniously.

From the perspective of ecological philosophy, the essence of ecological town is interpersonal harmony and harmonious existence between men and nature. Tangxia's construction of ecological town especially emphasizes that human beings are part of the nature, and that mankind's own development can only be realized on the basis of harmony and realization of overall coordination between mankind and nature. So integrity is the value that Tangxia Town sticks to in the constructuion of ecological town.

From the perspective of ecological economy, an ecological town not only needs a sustainable economy to provide corresponding production and living conditions to meet the basic needs of its inhabitants, but also has to ensure the quality of the economic growth. An ecological town advocates the promotion and popularization of green energies, commits itself to the efficient use of renewable energies and the prudent use and recycling of non-renewable energies, and pays close attention to the development and cultivation of human resources. Thus sustainability is Tantgxia's powerful impetus in its construction of ecological town.

From the perspective of ecological sociology, an ecological city is not simply the ecologicalization of nature, but the ecologicalization of human beings, that is, featured by the overall ecologicalization of education, science and technology, culture, ethics, law and system, a fair, equal, safe and comfortable social environment in which people protect ecological environment conscientiously to advance human beings' own development is to be established. So all-round development is Tangxia's fundamental pursuit in building ecological town.

From the perspective of geographical space, an ecological city is not an enclosed system, but a synthesis of society, economy and nature depending on a certain region. Geographically speaking, an ecological town is integration of town and countryside, that is, the town and its surroundings become integrated and form a complete whole in which the town and the countryside coexist and are mutually beneficial, thereby realizing regional sustainable development. So compatibility is the source of Tangxia's vitality in building ecological town.

Dear friends, to realize the harmonious coexistence between human beings and nature is both our common responsibility and our common pursuit. We will take our winning of the title "International Eco-Safety Demonstration Town" as a new departure, work hard to build Tangxia Town into a modern new city with harmony, happiness and beauty, and make greater contributions to beautifying and greening our homeland!

Wish the Assembly a great success!

Thank you!

Partners in Cost Efficient Crisis Management *

Otto Evjenth, Director of World View Global Media

BROADCASTERS
PARTNERS
IN
COST EFFICIENT
CRISIS MANAGEMENT

Broadcasters - partners for social development

2

---

* This article comes from the speaker's PPT.

AIBD

OUR EXPERIENCES TELL US THAT:

- THE PROVISION, HANDLING, AND FORMATTING OF INFORMATION BEFORE UNDER AND AFTER CRISIS ARE INTERDEPENDANTLY CRUCIAL TO THE QUALITY AND THE RESULT OF THE INTERVENTIONS TO BE DEPLOYED FOR INDIVIDUALS AND SOCIETIES
- SYSTEMS FOR EFFECTIVE DEPLOYMENT OF INFORMATION RESOURCES AND PROFESSIONAL DEVELOPMENT AND USAGE OF SUCH RESOURCES COULD BE IMPROVED

Broadcasters - partners for social development

3

AIBD

To adjust to the experiences we have initiated the:

- DEVELOPMENT OF AN INTERNATIONAL REPOSITORY OF QUALITY LOCALIZABLE INFORMATION RESOURCES
- ENHANCEMENT OF AFFORDABLE ACCESS TO ADEQUATE INFORMATION BY DEPLOYMENT OF BROADCAST TECHNOLOGIES (Push VOD)
- IMPROVEMENT OF PROFESSIONAL MEDIA SKILLS FOR MEDIASUPORTED INTERVENTIONS DURING CRISIS

Broadcasters - partners for social development

4

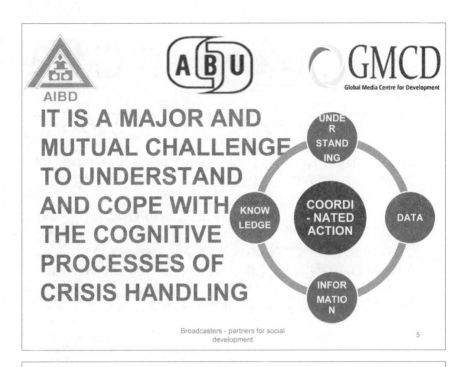

IT IS A MAJOR AND MUTUAL CHALLENGE TO UNDERSTAND AND COPE WITH THE COGNITIVE PROCESSES OF CRISIS HANDLING

UNDER STANDING

KNOW LEDGE

COORDI-NATED ACTION

DATA

INFOR MATION

Broadcasters - partners for social development

5

AIBD

To be able to get to and optimal model for information handling during crisis we:

1. will have to establish collaborate models under UNESCO coordination
2. will have to establish international platforms which are allowing and motivating for resource sharing
3. will have to further develop the role and the value of the "Community Information and Communication Centres" as a core units through all phases of crisis management.

Broadcasters - partners for social development

6

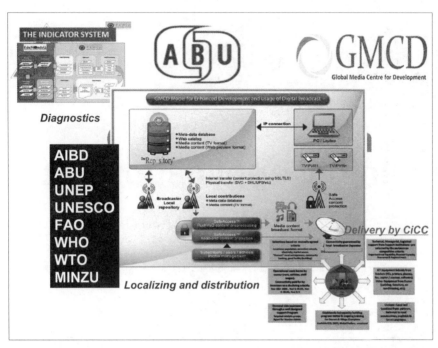

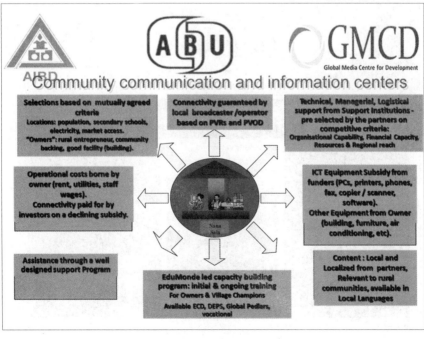

Initial applications at hand from broadcasters

9

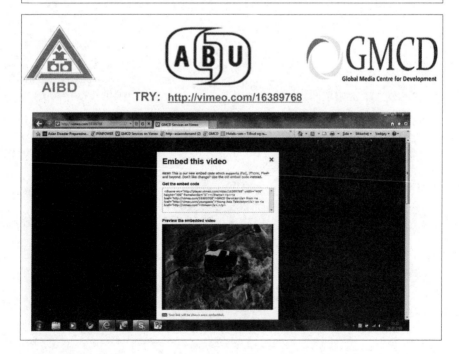

BROADCASTERS ARE READY TO
TAKE A POSITION AND TO PLAY A ROLE

12

IV
Thematic Session
(Eco-Safety: Strategic Vision of International

Multilateral Cooperation)

All Participation to Enhance Ecological Safety Education

Kol Pheng, Senior Minister of the Royal Government of Cambodia,

and Permanent Standing Member of FUNCINPEC Party

Excellency Chairman,
Honorable Delegates,
Ladies and Gentlemen,

Good afternoon!

I have the honor and privilege to contribute my humble understanding to this great Assembly on the critical issues of Eco-Safety that our world is facing today. We all know that the Earth is the only place on which human being can live. Therefore, when the living conditions on Earth are deteriorated, it would also mean that mankind faces a direct threat for survival. It has been inherited by the today world due to modern technology and the force of industrialization. It can be said that men are victims of their own inventions.

We are gathering today, trying to find solutions to hold back the swift degradation of our environment before mankind encounters more and more crises and calamities to the extent of a total collapse of the environment. The fact that the environment upon which we depend for survival is progressively collapsing because of human abuses. It currently becomes obvious that when industrialism, commercialism, and materialism becomes the mainstream economic model in which the whole society is focusing only on money making, man's mind loses its rationality to distinguish needs from wants. Hence, the result is that all pressures on resources pass on to our Mother Earth, causing tremendous damage to the natural ecology, thereby threatening man's very own survival. If the so-called abundant material life has our own survival as its price, then this is too high a price to pay.

Man is the smartest living creature on Earth; the technological and scientific advancement is recent years is the best proof of this. However, if there were only the advancement in science and technology alone without moral criteria, compassion to our fellow men and other sentient beings as well as nature, then such advancement may not necessarily bring about security and happiness for mankind.

Excellency Chairman, the answer to the above crises has to be sought from different angles, including science, philosophies and natural law. We should recall that Lao Tzu advised us that "men should live in harmony with nature". This simply means that men should not violate the natural law. Along with this, Mahatma Gandhi said "The Earth provides enough for human's needs but not enough for human's greed."

On this critical issue, the Buddha teaching on greed, moral degradation and its impacts on the nature can be used as an explanation of our current ecological problems. When king or ruler of a nation lacks of virtue, his wizards and officials at all levels, and his merchants as well as the general populace do not comply by the law and orders of the kingdom. When people reached this state of mind, they can do everything to meet their desire or greed, including violation of the natural law (the Dharma). Subsequently, the sun, the moon and the wind create a change in natural path, yielding unusual seasons. When this condition happens, the Mother Nature or what all of us called God gets angry. Then the rain does not come as usual. As consequences, the crops yielded under these unusual seasons and inappropriate rain will not provide people with good quality to sustain healthy conditions, thus shortening human life span.

On the contrary, when people abide by law and orders of the nation and respect the natural law, along with good virtue, the Mother Nature or God is definitely HAPPY. Hence, people will live in peace, happiness, prosperity and longevity. This is what Lord Buddha said: "One reaps what one sows".

Furthermore, the Buddha pointed out that the root sources of all evil actions are greed, hatred, and ignorance performed by sentient beings. It is a natural tendency of human being to have greed, selfishness, hatred and ignorance. However, we can tame our mind to overcome these evil thoughts, which hinder our true happiness in life.

Therefore, Buddhism cultivates a friendly attitude towards nature and looks at the relationship of plants, animals and people to one another from the angle of friendship and harmony. Also, we should recall the three major events in the life of Lord Buddha: birth, enlightenment and demise. All these events took place under trees in the open air. Lord Buddha advised His disciples to seek out forest for meditation. The delightful, serene atmosphere in a natural environment was considered as an incentive for spiritual growth.

Furthermore, compassion has always been the Buddhist ethical principle. Compassion refers not only to no-injury, not-harming the lives of other living beings but also to alleviating sufferings of others.

So far we can see ecological issues that we are confronting today are due to the unprecedented inflation of man's unlimited greed and moral degradation. The core of environmental crises resolution still lies in limiting an eradicating greed in human mind. We can do this by practicing loving-kindness, compassion, cultivating habits of frugality and simplicity, holding ourselves responsible towards ecological safety. Collective action is needed to reduce consumption, avoid wasting energy and polluting the air and water sources, and voluntarily give space and time to animals and plants to growth.

Our duties as leaders are to educate our citizen, our present and future generations to increase their moral value and to develop virtues in order to prevent moral degradation. We believe that we can work together on this special spiritual path, and then our environment can be rehabilitated. When we treat the nature nicely, the nature will treat us kindly in return.

The Earth cannot be replicated, therefore let each and every one of us cherish our planet, our Home.

May All Beings Be Always Happy!

Thank you so much for kind attention!

Work Jointly to Promote Cooperation in International Eco-Safety

Philippe Le Gall, Ambassador Extraordinary and Plenipotentiary of

Seychelles to China

Mr Chairman,

Excellencies,

Ladies and Gentlemen,

Good afternoon!

Please allow me first to express my thanks to the Government of the Kingdom of Cambodia for its warm hospitality and for all practical arrangements to make our stay in Phnom Penh both fruitful and pleasant.

Let me add special words of thanks to Professor Jiang Mingjun, Director General of IESCO, for having invited me to take the floor during this thematic session

Mr Chairman, This morning, my Foreign Minister made a presentation on Seychelles' pro active stance on climate change and ecological safety, and on a number of challenges related to the sustainability of oceans. So, from the start, I would like to renew our thanks to WESA organizing committee for this opportunity given to island states, especially small island developing states, to express their position on such important and urgent matters.

Mr Chairman, the subject of this session takes us into what is fortunately no longer an uncharted territory, I mean: international multilateral cooperation to implement at global level a strategic vision of ecological safety.

But still a lot has to be done and of course it may be difficult for politicians and policy planners to share a common strategic vision, when on one hand they consider the modest results achieved in Copenhagen and during post Copenhagen climate change negotiations till now, or due, on the other hand, to the often confusing diversity of views and scenarios reflected in the debates and publications of the Intergovernmental Panel on Climate Change, the Heartland Institute and MIT for instance, to name but a few institutions whose reports — once summarized into news — are frequently perceived by public opinions round the world as indulging in either doom-mongering or excessive optimism.

But the point remains that without such strategic vision of ecological safety and international multilateral cooperation the search for solutions to the global environmental degradation that is threatening our respective countries and regional economies will fail to produce encouraging results.

This is particularly true for small island developing states

Mr Chairman, Seychelles is a small country when you consider its land mass, less than 500 square kilometres, and its population, around 85.000. If I am not mistaken this is approximately the number of Cambodian-American living in the State of California alone or half the population of a city like Siem Reap !

But Seychelles is a big country when you consider its Exclusive Economic Zone of 1.3 million square km of ocean. Actually Seychelles ranks 24th, between South Africa and Mauritius, and this is a huge territory; it explains why my country is often compared to a small fish in a big aquarium.

However we cannot rely only on our limited human and natural resources to address all the challenges and threats we are facing, in connection with climatic alterations.

We need each other.

Mankind is a social species: yes, we do need each other and cannot live and prosper in seclusion. It is the same for countries. Countries are social entities and in an inter-dependent world they also need each other more and more as the 1st and most critical resolution threshold for many of these challenges and threats is at sub-regional and regional levels.

In this context, international multilateral cooperation provides useful and timely platforms to bring out, through consultations and dialogue, the consensual elements without which no vision can assume to be strategic.

Mr Chairman, today, I would like to highlight 3 channels of international multilateral cooperation in which more coordinated efforts have to be made to harmonize policies and action plans.

The first one is institutional.

I am referring here to regional political and economic partnerships. Being also Ambassador to China, Japan, Korea, and Vietnam, I would like to stress the importance of the programs of cooperation these countries have, in particular with Africa, and how crucial it is for member states joining these programs not to focus exclusively on poverty reduction, health, infrastructures and food security, but include ecological safety in their agenda as an additional and vital priority.

It is within such partnerships and at the level of regional and sub-regional organizations that much of the home work conducive to concrete results at the UN and in international forums is done so there is a constant need to engage into discussions at this particular level and agree on the best principles and practices of sustainable development we need to move forward within international organizations.

Mr Chairman, the second and the third channels are promoting smooth multilateral cooperation.

The second channel is through World Expos and International Expos.

I worked for the last two years with Shanghai World Expo as Commissioner for Seychelles pavilion and later on as a member of the Expo's Steering Committee.

As you know the theme of the Expo was "better city better life ". In Seychelles we only have one city for the time being, Victoria, one of the smallest capitals in the world, with less than 30.000 people live there. It would have been presumptuous to promote our city planning and urban development policy as a model for the rest of the 190 countries + 56 organisations having participated in the country...and consequently for the 73 million of Expo visitors.

But what we wanted to communicate to our visitors and share with them was a very simple message about the unique relationship we have with nature in Seychelles, about our belief that we do not own Nature, that we belong to Nature, and about the way we have translated this traditional belief into a national policy and a balance between economic and social development, natural heritage conservation and cultural heritage preservation.

We wanted our pavilion to be as nature-oriented as possible and serve as an eco-friendly wake-up call – if I may put it this way. And it worked very well ! 22 million people visited the African joint pavilion where our national pavilion was located. 22 million visitors ! I am not saying all of them got Seychelles' message but definitely a big percentage of them.

In 2012, Yeosu's International Expo, in Korea, on the theme " The living ocean and coast" will act in the same smooth way and I think that for all island states and coastal countries represented here today, as well as for IESCO, it is a must to participate in this Expo and promote our common agenda for a better planet.

Mr Chairman, in conclusion, the third channel of smooth multilateral cooperation I want to mention is IESCO itself which raises international awareness and understanding of what Climate change could soon or later mean to all our countries and also plays the role of catalyst in identifying and promoting scientific research and economic initiatives to ensure that we are better prepared to adapt to our changing world and to the economic and social risks we are facing as well as the opportunities such risks include.

I would therefore like to pay a special tribute to IESCO for its open spirit of collaboration with all our countries and its special interest in Seychelles.

Thank you for your attention!

Shared Goals of Kansas City and IESCO

Dennis Murphey, Chief Environmental Officer, Kansas City,

Missouri, USA

Attendees of the 1st World Eco-Safety Assembly and distinguished guests,

I want to thank Director-General Jiang Mingjun for inviting me to attend this 1st World Eco-Safety Assembly and, as the Chief Environmental Officer of the city of Kansas City, I'm honored to speak to you on behalf of my community.

Kansas City, Missouri is located in the center of the United States, in an area we often call the Heartland of America. It is located on the banks of one of America's great rivers – the Missouri River – approximately 200 miles from its confluence with the Mississippi River. In the 18th century, Kansas City was the point of departure for explorers, settlers, and traders who were moving into the Great Plains, the Rocky Mountains, and the desert areas of the American West.

In October of this year Kansas City was honored to host Director-General Jiang and other senior officials from China at the 14th Biennial Edgar Snow Symposium, an event that alternates between Beijing, China and Kansas City. This year's Symposium featured a Clean Energy & Green Solutions Summit where Director-General Jiang made an excellent presentation regarding the concept of eco-safety.

Although the term "eco-safety" was new to us in Kansas City, the concept and the elements of eco-safety are very familiar to us and represent goals we share with IESCO. Over the past several years, with the leadership of Mayor Mark Funkhouser and the twelve members of the City Council of Kansas City, climate protection and sustainability have been top priorities in our operation of city government. In Kansas City, when we say "sustainability" it means making decisions and taking actions that simultaneously promote the economic vitality, the environmental quality, and the social equity of our community.

In his presentation at the Clean Energy & Green Solutions Summit, Director-General Jiang stated that ecological disasters caused by global warming and human economic activities have posed a serious threat to human survival and national security, and the resulting frequency of disasters, loss of human life, and property loss is still increasing.

I completely agree with his conclusions and strongly share his concerns. Kansas City is one of a growing number of American cities who are taking action at the local level, in concert with our business community, concerned citizens, and other cities in our metropolitan area, to address climate change by reducting greenhouse gas emissions from municipal operations and all sources community-wide.

Through the U.S. Conference of Mayors, an organization representing cities across America, more than 1,000 mayors of cities have signed a climate protection agreement, committing them to assess and reduce greenhouse gas emissions in their communities.

Mayor Funkhouser is one of the signatories to the U.S. Conference of Mayors Climate Protection Agreement and, in July 2008 Mayor Funkhouser and the City Council of Kansas City unanimously adopted the Kansas City, Missouri Climate Protection Plan. This plan was developed in concert with a coalition of community leaders that represented neighborhoods, concerned citizens, environmental organizations, the business sector, local energy utilities, and our metropolitan area planning organization.

Kansas City's climate protection plan commits our city to reducing greenhouse gas emissions by 30% below year 2000 levels by 2020. The plan includes 55 specific greenhouse gas reduction measures to achieve that goal through improving the energy efficiency of our buildings, streetlights, traffic signals, and water/wastewater treatment facilities. It also calls for generating more of our electricity from low-carbon or carbon-free sources, such as wind turbines and solar panels. Other greenhouse gas reduction measures in our plan include reducing the fossil fuel consumption in our transportation system by expanding public transit and increasing the viability of walking and biking as transportation options. Our longer-term goal is to reduce our greenhouse gas emissions 80% below year 2000 levels by 2050.

These goals are ambitious, but the threats to national security and human survival on our planet from climate change are too great for anything less than achieving ambitious goals in greenhouse gas reductions. In America we have recognized that, in the absence of sufficient progress at the national level, individual local governments must take decisive, meaningful actions that cumulatively will contribute in a significant way to mitigation of adverse climate change and its consequences. Local governments also continue to call upon our national leaders to adopt policies to reduce greenhouse gas emissions and to work cooperatively with all members of the international community to address climate change in a meaningful way.

Therefore, in Kansas City we believe that IESCO's work to realize the U.N. Millenium Development Goals of eradicating extreme poverty and hunger, ensuring environmental sustainability, and developing global partnerships for development is critical. We applaud IESCO's efforts and congratulate them for convening this 1st World Eco-Safety Assembly.

With regard to global partnerships, I am currently working with a non-profit organization in the U.S. called the Institute for Sustainable Communities (ISC). As a member of ISC's China Climate Leadership Academy faculty, over the next two years I look forward to working with Chinese provinces and communities, non-government organizations, and universities to develop and implement climate protection strategies. Partnerships such as these are important ways that local governments in our countries can share information and work together to address the most challenging global environmental issue that we face, not individually, but collaboratively. An effective response to climate change is beyond the ability of any single country – its solution lies in the cumulative impacts of cooperative efforts by the entire community of nations. Local governments in countries across the world are stepping forward to provide the needed leadership in this area.

I am very honored to be here today to share with you some of the activities we have undertaken in Kansas City and to learn more from you regarding how we can do a better job of incorporating eco-safety principles and practices into our local government operations. We share the goal of aligning our decisions and actions in ways that will promote eco-safety and a positive legacy for future generations across the globe.

On behalf of Mayor Funkhouser, the City Council, and all the residents of Kansas City, I thank you for inviting us to participate in this important event and providing an opportunity for us to open a dialogue with IESCO that we hope will lead to additional ways we can cooperate with you in ensuring a better future for our world and all of its peoples.

Pay Attention to Activity Directions and Subject Study of Ecological Alliance

Karpush Malayan, Vice president of the International Academy of Ecology and Life Protection Sciences

Ladies and gentlemen, friends and colleagues,

Please allow me, on behalf of the International Academy of Ecology and Life Protection Sciences, to extend my heartfelt greetings to all the distinguished guests participating in the Assembly, and express my sincere thanks to the organizing committee for organizing and hosting the 1st World Eco-Safety Assembly on this land of hospitality.

Today we are gathered here, thanks to the great efforts of Mr. Jiang Mingjun and his team as well as the Royal Government of Cambodia, to whom I would like to express my deep respect and gratitude.

To begin with I would like to quote the late 18th century French scholar Lamarck: "To put it this way, the whole meaning of human beings is to destroy their own race, and to make Earth an inhospitable place."

It was in the distant past that he was already anticipating and defining today's social crises which were hard to imagine then – the crisis of technological civilization and the crisis of the complex relationship between human and nature. In the 20th century, we humans were still breathing fresh air, drinking clean water, and enjoying the seemingly boundless world and endless natural resources. Only decades later, the world has been on the verge of terrible anthropogenic ecological crises. If human beings continue this path, they will not escape the demise a few generations later. This is a view held by many ecologists and pessimists, both of whom might be realists by nature

What has caused the so-called ecological crisis? And what has lead to its emerging at the end of the 20th century and development in the present society? Some scholars believe that it is based on two factors: the population growth (the global population of the early 21st century has increased by six times compared to the early 20th century's level of about one billion) and the technological revolution which has greatly promoted the consumption of natural resources (energy consumption has increased by 10 times, resources usage has increased by 9 times, the growth rates of which have both exceeded population growth).

The concept of "Ecological Crisis" was first presented in 1972 in a speech of the Club of Rome, an organization committed to studying contemporary global issues under the leadership of Meadows. In the essay, the following conclusion was made: according to the trend and the accelerated rhythm of economic development, humans will meet their doom in 2100, when most humans will die from hunger and lack of resources, natural resources will not be enough for human survival and production in term of basic raw materials, and Earth will become inhospitable for humans due to environmental pollution.

Therefore, the contradiction between the human goal (improving living standards) and the natural conditions is now on the table. The destruction and intervention of nature by human will lead to the destruction of nature as well as human.

Thanks to the intellect and perseverance of all those advanced minds from the contract countries, in July 1992, United Nations Conference on Environment and Development was held in Rio de Janeiro, with the outcome documents of Agenda 21, a comprehensive plan of action to be taken globally, and Sustainable Development, an international strategic approach to Earth civilization development that was approved for the first time in world history, and which, according to some scholars, should have been put as — program for human survival, because the almost exhausted natural resources cannot serve as the backbone of economic growth. The following are the trends of the last century, the possible direction of activities and subjects for research summarized by our ecological allies — International Academy of Ecology and Life Protection Sciences, International Eco-Safety Cooperative Organization and other authoritative international organizations concerning about ecological safety:

The trends show crises in the following order of importance:

1. Global climate change;

2. Damage to the human organism;

3. Free use and quality of water;

4. Depletion of mineral energy resources;

5. Loss of biodiversity;

6. The destruction of the ozone layer;

7. Limitations of the features of land use

It is my sincere hope that in future speeches, the list would be more accurate and extensive, so that we can focus on the study into above issues and other issues not included with concentrated efforts and resources, and achieve certain results and answers.

I believe my colleague Appollonskiy will describe more accurately the fields where we can jointly work on the issue of the benefits of mankind in the framework of international conventions and other documents.

Today we attend the Awards Ceremony to award those political leaders, academics and entrepreneurs who have made important contributions in the field of ecology, and to whom I would like to extend my heartfelt congratulations and sincere wishes.

Only by joint efforts can we protect our planet from ecological disaster!

To conclude, I would like to quote the outstanding Russian scholar and philosopher Nikita N. Moiseev. In his final book, he proposed to call a society rational and organized, if this society has the ability to realize harmonious coexistence with nature and to coordinate its own development logic and that of nature. But Moiseev also put the existence of such a society as utopian. "Is it possible?" Moiseev asked this question and gave his own answer: "It is always hard to find the answer: yes or no." He concluded his book with the words below: "We have no time to form the moral basis to protect our common future ..."

Thank you for your patience and attention!

Subject Study and Prospect in the Field of International Eco-Safety

Stanislav Appollonskiy, Vice President of IAELPS

Respected ladies and gentlemen, dear distinguished guests,

I am glad to give my best regards to all of you on behalf of IAELPS at the First World Eco-Safety Assembly.

I understand what kind of difficulties the organizer of the Assembly has experienced and overcome. I'd like to express my profound appreciation and bow my thanks to Mr. Jiang Mingjun and his team, as well as all the representatives of Cambodian government who took part in the preparation and organization of the Assembly.

I hope that this pioneering work will make more people, like those participants of the Assembly, to pay a close attention to the problem of eco-safety which our humanity is facing, for it is no accidental that it is listed on the United Nations agenda.

It must be pointed out that the IAELPS has made the eco-safety a basic international work from the beginning of its establishment. The presidium IAELPS and its President, Rusak O. N, have closely followed the movements in the field of eco-safety, and insisted on the working policy of actively participating in the cooperation of international authoritative organizations.

After a great many organizing and research work, the IAELPS registered at UN News Center in 2000 and became a member of United Nations Economic and Social Council (ECOSOC) in 2003.

We had done a great deal of multifarious preparatory and technical work before the IAELPS was admitted into UN News Center and involved in their work. There are three basic objectives of IAELPS' going into the United Nations' institutions. The first one is to provide experts and scholars of IAELPS with access to the information of world scientific achievements to enhance its scientific level. The second is to let more people get to know the new technologic achievements of IAELPS in the field of ecology and life support and broaden their info view, with the help of organizations and institutions in the world. The last one is to make IAELPS participate in more international programs and projects under the support of United Nations and many funds.

From year 2004 to 2010, the agenda of IAELPS were in complete conformity with the following United Nations' program.

— Movement of nature conservation.

— Framework Convention on Climate Change.

— Protection and rational utilization of sources under the development objectives.

— Protection of the atmosphere.

— Chernobyl disaster and the development of nuclear power.

Scholars of IAELPS have made great contribution in the project research of nature conservation. They firstly put forward the division method of basic standards of ecological technology which has become popular. The most important and fundamental theory of this method is the concept of eco-activity coefficient based on the systematic analysis which distinguishes the ecological resources from the ecology in the process of manufacturing technology.

The experts of IAELPS have also made studies on some other important problems, for instance:

— The continuous increase of uncontrolled electromagnetic current and its pollution to air.

— Garbage pollution produced by manufacturing production, Dioxin, for instance.

It is estimated that these pollution will have an inestimable dangerous influence on humanity, plants and animals on the earth, and the radioactive contamination has the same extreme danger. But the experts of IAELPS have accumulated a wealth of theoretical and practical experience which will benefit countries all over the world.

The only way of solving the earth's eco-safety problem is through the cooperation of countries in the world, including the cooperation between developed and developing countries. So, we spared no effort in supporting Mr. Jiang Mingjun in his establishment of IESCO, an institute which tries to provide a solution to the eco-safety problem, protect ecological environment and realize the strategy of sustainable development. And soon after, out of the same reason, we set up the China Branch of IAELPS and elected Mr. Jiang Mingjun as its leader and the Vice President of IAELPS.

Time has proved that our choice is right, for the China Branch of IAELPS has achieved important results in improving national eco-safety. These results have been sufficiently reported in the International Ecology and Safety Magazine published by IESCO, and gained wide popularity.

The presidium of IAELPS set a high value on the results their Chinese colleagues have achieved, hoping that they will continuously consolidate and develop the cooperation forms in the future.

It seems to us that the projects in the following fields have broad prospects:

1. Perfect the theoretical foundation for those unresolved problems to make sure that effective decisions are adopted; promote sustainable development in which the effective utilization and the consumption of energy, transportation, industry, ocean resources and land utilization etc. are included.

The basic and ultimate goal of this project is to reduce the energy industry's discharge of hazardous substances to the atmosphere. The focus is the development of renewable and recyclable energy system, especially the utilization of nontraditional and recyclable energies, for instance, solar energy, tide energy and wind energy etc. The employment of more effective production, transportation, distribution and utilization method that can reduce pollution is also on the agenda. However, the goal must reflect the necessary fairness. We should consider the developing countries' essential and continuously increasing energy demand and the practical situation of some countries where the state revenue mainly comes from manufacturing production, procession and export, or others where traditionally, coal and petroleum are used as fuels. These countries' difficulties in their transition to the utilization of clean energy should be taken into consideration. In addition, the vulnerability of the countries where the climate has been changed by harmful activities should also be put into account.

2. Protect the stratospheric ozone from damage.

3. Transboundary Air Pollution, which includes the monitoring of radioactive and electromagnetic pollution, and the adoption of preventive measures, study of the

influence of radioactive and electromagnetic activity on the change of environmental indexes and the method of control, study of air pollution caused by heavy metal oxides discharged by industries, and study of earth surface pollution by industrial garbage.

4. Prevent predatory deforestation which will bring about disastrous consequences to humanity.

It must be pointed out that a deeper research into the project of eco-safety at the global level will make us sketch more clearly the way of safeguarding life and people's fitness by employing new technological achievements in a crisis.

In current society, however, the major orientation of safeguarding is the disasters caused by technology. According to the statistics of relevant departments of United Nations, the technical disasters occupy over 70% of all the emergences.

The main working directions of our experts of IAELPS include:

— Plan and utilize rationally the natural resources, such as land, water, energy and ecological resources and so on by using comprehensive measures.

— Promote sustainable development of rural economy and areas.

— The ecologically safe application of biological engineering technology.

— Preserve water system, and protect, utilize rationally and develop life resources in it.

— The protection and supply of fresh water resource; develop water resources with combined method and make an in-depth study on water economy and water utilization.

— The ecologically safe administration of the utilization of hazardous chemicals, including the prevention of illegal international traffic in toxic and dangerous products and harmful chemical waste; disposal of harmful garbage, hazardous radioactive waste and solid waste; and the purification of waste water.

— Fight with deforestation and desertification.

— Promote ecologically safe technology; participate in various cooperation and build up scientific power.

— Data statistics of environmental protection problems and in process of solving them.

— Set up a comprehensive statistical system of ecological and economic factors.

The experts and scholars' high level in the above-mentioned fields has given the IAELPS more confidence to perform in the future international activities. With persistent proposals from leaders of headquarters and regional branches, members of IAELPS have got the chance of applying their research results to practice and participated in the international operation, including projects invested by the United Nations and many international funds.

The presidium of IAELPS also showed its extreme concern for the work of Global Experts Evaluation Committee which is formed by senior experts of IAELPS and responsible for the evaluation and safeguarding of all important item designs. I hope there will be more experts who come from different countries and concern about problems of eco-safety to join our committee.

Today, we hold Commendation and Award Ceremony here to commend a group of politicians, experts, scholars and entrepreneurs who have made important contributions in the field of eco-safety. This commendation, however, is only a new departure in our dedicating joint efforts to the cause of eco-safety and sustainable development at a higher level, with the frequent natural disasters and ecological crises that face us, what we have done is not nearly enough and a great deal remains to be done.

Finally, it must be pointed out that the one-time event, even such important as the First World Eco-Safety Assembly, cannot solve all the eco-safety problems that face humanity, and concerted efforts of people and national governments in the world is required. So, in order to solve the eco-safety problems, we must do day-to-day meticulous and complicated jobs. Besides, we should not ignore the reality that the increasing eco-safety problems caused by the development of industry are badly in need of solution.

Ecological Safety: Strengthen and Deepen NGO Participation

Gennady Shlapunov, Coordinator of the Russian National

Committee of UNEP

Honorable Chairperson, ladies and gentlemen,

It is my pleasure to be in one of the most beautiful capitals in Asia, Phnom Penh, Cambodia. It is a great honor for me to deliver a speech in front of all honorable guests. I listen attentively to my co-workers' presentations this morning. All of their speeches, regardless about theoretical issues or practical issues, are all rich in context. Let me now briefly introduce my thought on the subject of eco-safety cooperation.

As early as 1996, Dr Jiang Ming Jun had already put forward the concept of eco-safety in an important international forum held in Vladivostock, Russia. He stressed that "in the 21st century, ecological safety, resources security, and sustainable development will capture more important position in international politics."

After 15 years, I can firmly say that prediction is coming true in certain domains. Those environmental issues steadily capture more important strategic position in foreign policy. Those political figures, social activists, academicians who put effort on eco-safety are now gaining more influence worldwide.

Eco safety is an important part of national security, and a key to global security.

Russia, as one of the great nations, needs to cooperate with other nations and takes up its responsibility in the field of eco- safety.

We cooperate with different international organizations and academic institutions, like Russian National Committee of UNEP, International Eco-Safety Cooperation organization, International Academy of Ecology and Life Protection Sciences, and International University of Business-technologies etc. By those cross-sector and multilateral cooperation, we exchange new idea and new techniques in the field of eco-safety.

I believe NGO has an important role in international cooperation and monitoring governments. Through the close cooperation with NGOs and academic institutions, we can deepen and widen international cooperation and monitor central and local governments' measure on eco-safety.

Russia — China borderline is roughly 4300km long. Without a doubt, eco-safety is a key issue in such a long borderline.

For example, Amur River marks the border between the China and Russia, roughly 3000km which has been polluted badly, and in a critical condition.

According to media reports, China's Helongjiang province alone discharges more than billion tons untreated domestic and industrial waste water into Amur River, which is higher than wastewater discharge standards in Primorsky Krai. Both sides, Russia and China, are also planning to build hydro-power project. Certainly, this project needs a high level cooperation between two governments, and needs to study seriously.

Although Russia and China inspect water quality of Amur River regularly, we, as a NGO, need to make a push for Russia and China to sign "The Convention on the Protection and Use of Transboundary of Watercourse."

Another important aspect of international cooperation is to maintain and preserve UNESCO World Heritage site.

As we all known, Lake Baikal is the deepest lake, contains the most world's surface freshwater, and has unique ecosystem.

As early as September 1994, Ulan Ude held International Conference on "Lake Baikal Region: A Model Territory for the World", which contributed a lot to Lake Baikal researches.

The World Bank initiated scientific research on Ecotourism Planning and Development in Lake Baikal region. The basic goal of this plan is to minimize the damage of natural environment, and to maximize the economic benefit.

And "Natural Resource Management and Biodiversity Conservation in the Basin of the Lake Baikal" sponsored by the World Bank also completed successfully.

Within the framework of "A Comprehensive Program of Land Use Policies in the Russian Territory of Lake Baikal" (Project Davis) between Russia and USA, a fact-finding research also was taken.

The aim of EU's TACIS program in Lake Baikal region is to preserve the natural environment. There are 2 advantages of this program: program operators can understand different European organizations from different directions and gain the practical experience on natural resource and environment quality management.

Those examples have proven that we are working on perfection to preserve Lake Baikal, and the prospect for further cooperation is bright and promising. However, we haven't given foreign investors an access to our ecological information when it comes to particular program operation.

As some Russian organizations' unenthusiastic attitude on cooperation, we have no choice but to delay the program. More specially, those organizations capable of coordinating consuming natural resources and protecting environment in Lake Baikal region are lack of initiative to cooperate.

In a global sense, Lake Baikal and its ecological system mean a lot, and it is international community's responsibility to preserve Lake Baikal. A lot of academicians with advanced thinking will agree my view. I firmly believe that international community will expand their cooperation in resolving ecological issues in Lake Baikal.

Those assemblies will be sponsored and organized by International Eco-safety Cooperative Organization. So far, many countries from Africa, Asia, Europe and Americas have expressed their interests in attending the Assembly.

I believe that I don't have to tell how critical of the ageing problem in modern society, I just want to make one point; the aging problem is closely related to ecological issues.

My co-workers and I have already begun to organize the assembly. It could be expected that our cooperation will continue to expand, and once again to bring a fruitful result from the assembly.

Thank You.

Power of the Youth in Global Sustainable Development

Shan Fengping, Chairman of United Nations Initiative and

Technology for the Youth (Asia-Pacific Region)

Honorable guests, ladies and gentlemen,

Good morning!

I am very glad to get together with you in Phnom Penh, Cambodia, and deliver my keynote speech in the name of United Nations Initiative and Technology for the Youth. The subject of my speech is "Power of the Youth in Global Sustainable Development".

The United Nations has realized long time ago that the imagination, ideal and energy of young people are crucial to the sustainable development of the society they live in. Based on this recognition, the UN member states signed Declaration on the Promotion among Youth of the Ideals of Peace, Mutual Respect and Understanding between Peoples in 1965. In 1995, when cerebrating the 10th anniversary of International Youth Year, the United Nations required the international community to continue its rapid reaction to the challenges the next millennium would pose to the youth so as to emphasize its commitment to young people. Therefore, it adopts an international strategy, that is, World Programme of Action for Youth towards the Year 2000 and beyond, (World Programme of Action for Youth for short), in order to more effectively cope with the problems young people have met and increase their opportunity of participating in social activities.

World Programme of Action for Youth is a specific action plan, which includes 10 areas: education, employment, hunger, poverty, environment, drug abuse, juvenile delinquency, leisure activities, girls and young women, and full and effective participation of young people in social life and decision-making. In each priority area, Programme of Action examines the nature of the challenges, and gives advice for action. These ten priority areas are interrelated and inseparable. For example, juvenile delinquency and drug abuse are often directly caused by inadequate access to education, employment and participation in social activities.

Under the above-mentioned background, countries such as India and Bangladesh initiated and established the international organization—United Nations Initiative and Technology for the Youth, which was approved and recognized by the United Nations General Assembly Special Session in 2001. Based in Kolkata, India, UNITY is affiliated to the Second Department of UN-HABITAT. On the basis of World Programme of Action for Youth, UNITY aims to lay stress on the rightful status and development as well as the living environment of women, empower the youth to action and enable them to create a living environment that is sustainable, politically and economically feasible, and socially just at present as well as in the future.

In order to call more attention to the problems and challenges young people are facing, UNITY works on a broad range of issues that have been identified as central to the livelihoods of women: eradication of poverty and marginalization, employment, health, human rights, education, environmental protection, prevention of Aids and women's grass-roots work in initiating and promoting sustainable development and environmental problems.

It is well known that young people are the main strength to push forward urban development. They are both the main human resources for development and the main promoters of social reform, economic development and technological renovation. Their imagination, ideal, abundant energy and longing are crucial to the sustainable development of the regions they live in.

Though the youth of the world live in countries with different phases of development and different social and economic environment, they all long for full participation in social life. However, the global society, economic and political situation that have undergone great changes make it very hard for young people in many countries to participate in social life. That is why Programme of Action should be carried out.

Therefore, UNITY attaches the uttermost importance to the development of education. In recent years, though every country has made prominent progress in popularizing primary education and eliminating illiteracy, the number of illiterate persons is still increasing. To encourage and develop the educational and training system that is more suitable for the present and future needs of the youth and the society, UNITY helps the youth in the developing countries gain more opportunities in college and higher education, research and vocational training. In light of the economic problems and inadequate international assistance these countries are facing, UNITY, together with governments of countries and other non-governmental organizations, helps the youth in the developing countries receive education and training of different levels in the developed countries and the developing countries, and conduct academic exchange among the developing countries.

Apart from education, employment is also one important work of UNITY. Nowadays, unemployment and inadequate employment of young people can be seen everywhere. It is very difficult for young people to find suitable careers. Moreover, they also encounter many other problems, including illiteracy and inadequate training, economic slowdown and overall transformation of economic situation, all of which make the situation very grievous for young people.

To solve these problems, UNITY, governments of countries and other non-governmental organizations formulate and promote donation plan, set up model co-op for the ones young people in the developed countries and the developing countries operate, give specific employment opportunities to specific youth group, coordinate governments of every country to appropriate resources from the funds, which are used to increase job opportunities for the youth, for programs which aims to support young women, disabled youth, demobilized youth, migrant youth, refugee youth, the homeless, street children and aboriginal youth, encourage governments of every country, especially governments of the developed countries, to create employment opportunities for the youth in the fields which changes rapidly due to technological renovation.

Comparing with education and employment, hunger and poverty are more fundamental problems. Today, there are 1 billion people in the world who still live in unacceptable poverty. Most of them are people in the developing countries, especially in the rural area of low-income countries and the least developed countries in Asia, the Pacific Ocean, Africa, Latin America and Caribbean.

To solve hunger and poverty, UNITY, together with governments of the various countries in the world, enhances educational and cultural services in impoverished and backward rural areas, takes other incentive measures to appeal more to the youth;

promotes experimental farming programs that are intended for young people, and expands extension service to the aspects of agricultural production and marketing so as to improve it; organizes and cooperates with local youth in different places to hold cultural festivals to strengthen the exchange between urban youth and rural youth; encourages and assists youth organizations to hold small and large-sized meetings in rural areas, especially the cooperation of rural population including rural youth; caries out training program according to rural economic demands and rural youths' needs for development of production and realization of food security. In this type of training program, UNITY specially attaches importance to skill training and ability promotion of young women, young people who stay at rural areas, young people who return from cities to rural areas, disabled youth, refugee youth and migrant youth, the homeless, street children, aboriginal youth, demobilized youth, as well as the youth living in the places where conflicts have been solved.

At present, UNITY is starting to build an overall action mechanism, which will continuously adjust necessary human resource and political, economic, social and cultural resources. However, it is governments of every country that are ultimately responsible for the implementation of this action mechanism and gain support from private departments and other social organizations. As long as we mobilize the strength of all walks of life to support the action plan of young people, I believe, they will certainly play a greater part in urban sustainable development.

Thank you!

Features and Advantages of DS-Circular Economy Series Technology

Shi Hanxiang, Chief Scientist of Beijing Hanxiang Research Institute

of Environmental Biological Sciences

Distinguished Guests, Ladies and Gentlemen,

Good afternoon! I am from Beijing Hanxiang Research Institute of Environmental Biological Sciences, and my name is Shi Hanxiang.

At the Award Ceremony hosted by International Academy of Ecology and Life Protection Sciences just then, I was awarded the title of Senior Researcher, for which I am very honored and excited. I would like to express my sincere appreciation to president of the Academy for the evaluation and recognition. Taking this opportunity, in front of all the leaders, experts and scholars present here, I would like to give a brief introduction on the "DS-circular economy technology series" developed by my team and me in recent years, and to share views with you.

With the rapid industrial development and accelerated population growth, large amounts of chemical substances produced by burning of coal and oil are emitted into the atmosphere in the form of waste gas and dust, which has exceeded the capacity of the atmospheric environment, caused frequent ecological disasters in today's world, and has posed a serious threat to human survival, resulting in a situation which is extremely serious. Industrial pollution caused by large amounts of gas and dust emissions in particular has become the chief culprit of global air pollution.

At present, in dealing with flue-gas pollution, most countries of the world have focused on desulfurization technology, and 90% of them adopted the limestone / gypsum desulfurization approach, but its drawback is that most of the desulfurization products produced after the limestone absorbs sulfur dioxide can not enter the recycling, which takes up large areas of land for stacking and forms secondary pollution.

The core of "DS-circular economy technology series" is that, with the solution of industrial three-dimensional pollution as the starting point, solid wastes such as steel slag and fly ash from industries such as steel and thermal power are used as sulphur dioxide absorbent to realize efficient desulfurization through the multi-phase reactor (equipment for this series of technology), and the desulfurization by-products are used for the improvement of saline-alkaline desertified land. It paved a brand new way for the three difficulties: the addressing of sulphur dioxide air pollution, the comprehensive transformation of slag and other pollutants and the improvement of saline-alkaline desertified land. It also realizes the resource feedback model of resources, products and renewable resources. While "transforming waste to resources with reduced volume and harm", it gradually extends to the recycling of resources and forms a complete industrial chain of circular economy.

"DS-circular economy technology series" has a strong adaptability to be used in all industries producing sulphur dioxide gas, such as coal-fired power plants, iron ore sintering, coal-fired furnaces, non-ferrous metal smelting, chemicals, and building materials; As for choosing desulfurization agent, industrial alkaline materials can be adopted to greatly reduce operating costs, such as non-ferrous metal smelting slag, iron-making slag, steel-making slag, carbide slag, saline-alkaline soil and metal oxides (such as zinc oxide, manganese oxide and magnesium oxide), according to the object of flue gas desulfurization and its condition, as well as local conditions.

The main features of the technology include: (1) the use of industrial slags such as copper slag, blast furnace slag, steel slag, red mud, white mud, carbide slag, fly ash, zinc oxide dust, blast furnace dust and recycled water from water slag, as desulfurization agent; (2) the use of standardized, modular and non-nozzle structure design; (3) the use of special high temperature polymer engineering plastics and overall molding technology, with advantages such as no scaling, no clogging, good performance in corrosion and wear resistance, which have practically realized high desulfurization efficiency, saving in investment, low operation cost and a long service life.

The most significant advantage of this technology is that: the SO_2-absorbed final product can be processed into soil conditioner and saline-alkaline soil improver, with heavy metal content in desulfurization products ensured in line with national standards on soil improvement, through the mechanism change of both the exchange between calcium and iron ions as well as hydrated silicide and sodium, potassium, magnesium ions in the soil to obtain washable ions; and the reaction between acids in desulfurization products and the salt of weak acid and strong base in the soil to produce neutral insoluble material and washable sulfate. In addition, there is sulfur, phosphorus, iron and other nutrients needed by soil in the slag desulfurization products, to improve saline-alkaline desertified land, promote the overall development of agriculture, to form a industrial chain of circular economy which extends from industry to agriculture, thus has successfully resolved the world difficulties in the reasonable disposal and elimination of desulfurization by-products confronted by all the previous desulfurization technologies, which truly complies with environmental laws and maintains ecological balance.

In recent years, during application and continuous improvement the circular economy technology has included 17 invention patents, 5 new model utility patents, and has been successfully applied in dozens of large and medium sized enterprises in such industries as non-ferrous or ferrous metallurgy, chemical industry and electric power, from home provinces such as Zhejiang, Hebei, Tianjin, and abroad such as Vietnam.

In my opinion, in the treatment of environmental pollution, we cannot walk the old path of "pollution first, treatment later". Instead, the traditional development model at the expense of environment should be overhauled, and economy development should be guided with the essence of Chinese culture and a dialectical thinking. As the citizens of the Earth, we should take on the responsibility and obligation of protecting our home planet, and make joint efforts in building a prosperous future.

Thank you!

Finally, let me finish by wishing the 1st World Eco-Safety Assembly a great success!

Features and Values of the Ecological New Energy Hai Ti Fu

Deng Chubai, Expert on Urban Sewage Treatment in China

All the respected experts, ladies and gentlemen,

Hello! I'm very glad to come to the beautiful Phnom Penh and to share the achievements of HTF ecological energy together with the political heavyweights, experts, scholars from various countries in the world. At this moment, I am very excited. I appreciate Vice Chairman Jiang Mingjun's recommendation and UN/ International Academy of Ecology and Life Protection Sciences' approval of my work. Taking this opportunity, I'd like to briefly introduce the issue process and major characteristics of HTF new ecological energy and I hope the honored guests present could benefit me with your precious suggestions.

While IT industry is being considered as the one with the most commercial value, and new energy industry with the most promising commercial prospect, "Microbe-Capturing-Carbon Recycling Fuel•Hai Ti Fu" is a new energy with both highest social and commercial value in 21st century.

In his new book The Ten Technologies to Save the Planet, Chris Goodall, a famous climate change expert from Britain, claims that ten technologies, if used by human, can prevent the planet from destructive climate catastrophe. And "carbon capturing" is one of them. "The increase of renewable energy has lagged behind the global demand for energy, which presents a great challenge to human that is to find an effective way to capture and store carbon dioxide from power plant. And carbon capturing comes as the solution," writes Chris Goodall. Although the related research goes at a slow pace, governments have realized the importance, which will give birth to new technologies.

As high-tech ecological energy, HTF is different from traditional bio-fuels energy, nuclear energy or wind energy. It is an effective and safe ecological fuel produced through mixed fermentation and biochemical synthesis of many microbial communities, such as photosynthetic microbe.

Firstly, the raw materials for HTF are castoff from the production and human beings' everyday life. And wherever human beings exist, castoffs will emerge. Today we are bothering to consider how to deal with these castoffs. While the production of HTF can transform the sewage into biological fuel, which can simultaneously reduce pollution sources and produce new energy. Therefore, it is the best practice in energy saving and pollution reduction.

Secondly, HTF is one of those fuels that consume the least energy during their production. HTF is a kind of carbon-recycling fuel synthesized by microbes under natural temperature and normal pressure. Compared with the production of alcohol fuel, HTF is more energy-saving, because the process of synthesizing wood fibers and starch into alcohol fuel needs saccharification and alcoholization under the conditions of certain temperature and pressure. While 'high-energy fuel' is synthesized under the normal temperature and pressure, and its main raw materials are industrial and urban castoffs and carbon dioxide. Therefore HTF is cheaper and safer than other biofuels fuels in terms of cost of raw materials and production.

Thirdly, HTF is the safest fuel with flash point of 29℃, it is very safe in term of application, transportation and storage. And we adopt the sterilization and disinfection process in its production and apply chelation to heavy metals, such as cuprum, to avoid the harmful gases possibly generated from the combustion of this fuel. After a few trials, burning HTF does not produce harmful gases, or black smoke in the circumstances of naturally mixed surface combustion.

Fourthly, HTF helps reduce greenhouse gas – CO_2. Presently all fuels give off heat through the oxidation of carbon, and the carbon in fuels will finally turn into CO_2 and be emitted into the air. From this perspective, heat generation will inevitably increase the CO_2 content in the air. But in the synthesis of HTF, CO_2 in the air has been captured and transformed into hydrogenous combustion particles, which can fully activate the heat reservation in hydrogen. It means that when the same heat value is generated, HTF will emit much less CO_2 than other fuels. Thus, the use of HTF will greatly reduce the CO_2 content in the air. And HTF is the safest fuel to control global warming and reduce CO_2 emission.

From the above four aspects, we can proudly say that HTF is the most ecological and the safest fuel in the 21st century and has immeasurable social and commercial values once goes into service. On June 2009, we have held the news conference in Beijing to officially popularize the use of this technology. At present, it has come into the mature probationary period and I really hope governments of all countries could attach importance to this technology and realize its value throughout the whole world.

Thank you!

V
Appendix

appendix

Appendix I : *

Confirmation Letter from the 6th General Assembly of ICAPP

ICAPP

ACKNOWLEDGEMENT AND ACCEPTANCE

To: The Organizing Committee of 6th International Conference of Asian Political Parties   (ICAPP) General Assembly

According to the announcement made during the 13th Meeting of the ICAPP Standing Committee at Kunming, Yunnan Province of China, Please be informed that we hereby agreed and accepted the proposal of the International Eco-Safety Cooperative Organization (IESCO) to include the 1st World Eco Summit (WES) Conference as part of the parallel 4 days program of the 6th ICAPP General Assembly in Phnom Penh, Cambodia, in the light of our urgent common focus on the interlinked battles against poverty, climate change and environmental degration. This is in furtherance to the suggestion of H.E. Sok An, Deputy Prime Minister of Cambodia and Chairman of the Organizing Committee of ICAPP's 6th General Assembly.

We concurrently accepted IESCO's proposal to invite the ICAPP members who might wish to travel to Siem Reap and stay overnight for their visit to Cambodia's Angkor Wat Heritage Area, which is one of the wonders of the world.

Dated 29th August 2010 at Manila.

Thank you.

Yours sincerely,

Former Speaker Jose de Venecia Jr
Founding Chairman of ICAPP &
Co-Chairman of the ICAPP Standing Committee

Cc: Secretariat of 6th ICAPP General Assembly
    Secretariat of ICAPP
    Secretariat of IESCO

---

*   This appendix is a scanned document.

Appendix II : *

Confirmation Letter from Government of the Kingdom of Cambodia

គណបក្សប្រជាជនកម្ពុជា
គណៈកម្មាធិការកណ្ដាល

Cambodian People's Party
Central Committee

៦គណៈ សន្និការ សេរីការ ប្រជាធិបតេយ្យ
អព្យាក្រឹត និង បូរណភាពសង្គម

Phnom Penh, September 09 , 2010

Prof. Dr. Jiang Mingjun
Director General
International Eco-Safety Cooperative Organization
<u>*Hong Kong, China*</u>

Dear Dr. Jiang Mingjun,

With reference to the letter of Honorable José de Venecia Jr., Founding Chairman of the International Conference of Asian Political Parties (ICAPP) and Co-Chairman of the ICAPP Standing Committee, dated August 29, 2010 I have the honor to confirm our agreement to include the First World Eco-Safety Summit in the program of the 6th General Assembly of the ICAPP scheduled to be held in Phnom Penh from December 1 to 4, 2010. For your information and reference, please find attached herewith a copy of the tentative program of the 6th General Assembly of the ICAPP in which you will find that the program for World Eco-Safety Summit is tentatively slated for Friday morning, December 3, 2010.

Taking this opportunity, on behalf of the Organizing Committee of the 6th General Assembly of the ICAPP, I would like to extend our invitation to the participants of the World Eco-Safety Summit for attending the General Assembly and its sideline events.

In this respect, I would like to suggest that you and/or your representatives to promptly coordinate the program of the Summit, protocol arrangements, list of the Summit's participants and the required budget with the Secretariat of the Organizing Committee of the 6th General Assembly of the ICAPP.

Please accept, Your Excellency, the assurances of my highest consideration.

SOK AN
Chairman of the Organizing Committee of
the 6th General Assembly of the ICAPP
and Deputy Prime Minister and Member of
the Standing Committee of the Cambodian People's Party

Cc: - Hon. José de Venecia Jr., Founding Chairman of the ICAPP and Co-Chairman of the ICAPP Standing Committee
- ICAPP Secretariat

---

*   This appendix is a scanned document.

Appendix III : *

Supporting Letter from United Nations University

UNITED NATIONS
UNIVERSITY
Centre

The 1st World Eco-Safety Assembly
The organizing Committee,

November 5, 2010

On the occasion of the forthcoming of the 1st World Eco-Safety Assembly, I would like to extend my congratulations on the successful convening of the Assembly on behalf of the United Nations University.

With advent of the 21st century, all kinds of natural disasters and sudden ecological catastrophes are breaking out at an increasingly higher frequency, and human society is confronted with a serious challenge from such ecological crises as global warming, desertification, water scarcity and pollution, destruction of forest and vegetation, loss of bio-diversity as well as problems in food safety and crops security, which has given rise to ecological safety as a main theme of the era with great impact on global development.

The 1st World Eco-Safety Assembly will provide a platform to share experience and search for solutions for governments, political parties, parliaments, non-government organizations, financial institutions, enterprise groups as well as experts and scholars from all corners of the world, which will help to open up social practices and close cooperation at different levels, and establish a global ecological safety cooperation mechanism.

As the think tank for the entire United Nations, United Nations University has, since its founding, been committed to promoting international academic exchange and cooperation, carrying out problem-oriented and multi-disciplinary studies of global affairs, and helping to resolve global issues related to human survival, development and welfare which are concerned by the United Nations and its Member States.

United Nations University is willing to positively support the convening of the 1st World Eco-Safety Assembly, and is looking forward to the great achievements of the Assembly.

Wish the Assembly a great success!

Kazuhiko Takeuchi
Vice-Rector of United Nations University

---

* This appendix is a scanned document.

Appendix IV : *

China Foundation for International Studies

Z.Y. H. 2010 (S.A.) No. 009

Secretariat of Organizing Committee of World Eco-Safety Assembly:

In charge of by the Chinese Foreign Ministry, China Foundation for International Studies (CFIS) is a national corporate body with members of senior Chinese diplomats, famous experts on international affairs and scholars. The aims of CFIS are to promote strategic and prospective study of the Chinese academic and economic circle on major international issues, and to develop academic exchanges and international cooperation.

And here we are pleased to inform you that we have received your letter that the International Eco-Safety Cooperative Organization (IESCO), the Royal Government of Cambodia and International Conference of Asian Political Parties (ICAPP) have scheduled to sponsor the 1st World Eco-Safety Assembly on December 1-4, 2010 in Cambodia. Considering that IESCO and our foundation are strategic cooperative institutes, we have agreed to co-sponsor the 1st World Eco-Safety Assembly.

China Foundation for International Studies

October 15, 2010

---

* This appendix comes from scans of documents on the same page.

Appendix V :

International Conference of Asian Political Parties Phnom Penh Declaration

Adopted at the 6th General Assembly of the ICAPP

held in Phnom Penh, Kingdom of Cambodia, on December 1-5, 2010

We — the leaders and representatives of (89) political parties from Afghanistan, Armenia, Australia, Azerbaijan, Bahrain, Bangladesh, Bhutan, Cambodia, China, DPR Korea, East Timor, India, Indonesia, Iran, Iraq, Israel, Japan, Kazakhstan, Kyrgyzstan, Laos, Lebanon, Malaysia, Maldives, Mongolia, Nepal, Pakistan, Papua New Guinea, Philippines, Republic of Korea, Russia, Solomon Islands, Sri Lanka, Syria, Thailand, Tonga, and Vietnam — gathered here in Phnom Penh, the capital city of the Kingdom of Cambodia, for the Sixth General Assembly of the International Conference of Asian Political Parties (ICAPP). The Sixth General Assembly of the ICAPP and the First World Eco-Safety Assembly were graciously hosted by the Cambodian People's Party, in collaboration with the FUNCINPEC Party, and International Eco-Safety Cooperative Organization from December 1 to 4, 2010, under the main theme of "Asia's Quest for a Better Tomorrow."

Noting that our Assembly coincides not only with the tenth anniversary of ICAPP's founding but also with the 32nd anniversary of the establishment of a 'Salvation Front for the Liberation of the Cambodian Nation from the Genocidal Regime,' we declare as follows:

ICAPP as an Open Forum for Asia's Political Parties

We reaffirm that ICAPP is an open and unique forum for Asia's political parties, and that it has become the pivot of inter-party dialogue and cooperation, working to achieve our common goal of sustained peace and shared prosperity in Asia. ICAPP is leading the way forward in what is generally accepted as the 21st Asian Century.

We note with great satisfaction and pride that, during its first decade, ICAPP has promoted exchanges and cooperation between political parties subscribing to competing ideologies; enhanced mutual understanding and trust among our peoples and countries; and promoted regional cooperation.

We reaffirm our commitment to the principles — and the spirit — of the ICAPP Charter and of ICAPP's subsequent Declarations at its biennial assemblies: in Manila in 2000, Bangkok in 2002, Beijing in 2004, Seoul in 2006 and Astana in 2009.

We reiterate our commitment to the intent and spirit of the United Nations Charter, International Law, the Five Principles of Peaceful Co-existence, and the Bandung Principles that emphasize democracy, good governance, human security, human rights, dignity, freedom, well-being, the rule of law and coexistence between ethnicities, cultures and faiths. Promotion of inter-faith harmony is basic tenet of ICAPP's principles. ICAPP embodies the Asian spirit of resilience evident in the dynamism and vibrance of our societies: successfully combating crises and overcoming the economic difficulties with innovation.

The Principles to Which We Adhere

We pledge to ensure peace, security, stability and prosperity for our home continent, in the context of growing political and economic multi-polarity, by adhering to the following principles:

— Sovereignty and territorial integrity of every state;

— Right to determine its own political, economic and social system of every state;

— Non-aggression and non-interference in each other's internal affairs;

— Peaceful settlement of territorial disputes and adherence to international treaties and laws;

— Arms control, disarmament and non-proliferation of weapons of mass destruction; and

— Rejection of every kind of extremism, prejudice or bigotry.

We support in particular the international agreement that "all States need to make special efforts to establish the necessary framework to achieve and maintain a world without nuclear weapons," as stated in the final Document unanimously adopted in 2010 Review Conference of the Parties to the Treaty on the Non-Proliferation of Nuclear Weapons, as well as the United Nations Secretary General's five point proposal which includes negotiations on a nuclear weapons convention.

We support our Standing Committee's initiative — in company with the Permanent Conference of the Political Parties of Latin America and the Caribbean (COPPPAL) — to reach out to our counterparts in Africa and in other continents to enhance mutual understanding and cooperation through dialogue and exchange, with the object of eventually convening a global convention of political parties. This was reaffirm at the meeting between Standing Committee of the ICAPP and Coordinating Body of the COPPPAL.

Facing up to Our Gravest Threats

We realize that environmental degradation and poverty passed down from generation to generation are the gravest threats confronting humankind in our time. We support all efforts — international and national — to moderate the effects of climate change and to alleviate hunger, ignorance and ill-health. We endorse U.N. Secretary General Ban Ki-moon's efforts to accelerate the attainment of the U.N.'s Millennium Development Goals; and to facilitate international agreements to reduce greenhouse gases. We endorse ICAPP's initiative to gain observer status to the General Assembly of UN, so as to coordinate its activities with relevant U.N. programs.

We realize the Asian economies' need to improve their management of financial liquidity and their ability to pool resources, if they are to control the processes of national and regional development. Our countries must seek inclusive growth by expanding economic parity and by increasing their share of the benefits from global economic growth. They can achieve this by bridging the development gap and by speeding up their economic integration through the connectivity being generated by globalization.

Our Call for Open Regionalism

We realize our need to boost intra-regional trade through open regionalism — by eliminating both tariff and non-tariff barriers, deepening cooperation, and integrating the economies of the Asian regions. We must link up the ASEAN and SAARC (South

Asia Association for Regional Cooperation) frameworks; strengthen the Mekong-Indo-China economic corridors; connect the Greater Mekong Subregion (GMS) with the South Asia Sub-regional Economic Cooperation (SASEC) as well as with the Asia Pacific Economic Cooperation (APEC); and bring Gulf Cooperation Council (GCC), Central Asia Regional Economic Cooperation (CAREC) and all other regional cooperation frameworks into a larger pan-Asian architecture.

In this regard, we remind ourselves of our vision and our ultimate goal to build an Asian community that will bring about shared prosperity to all peoples in the region and emphasize the importance of collectively striving to further strengthen coordination and cooperation among all countries and regional groupings, in particular through expanded roles of political parties.

We reaffirm our resolve to establish an Asian Anti-Poverty Fund and an Asian Micro-Financing Fund — as called for by the Kunming Declaration of the ICAPP Conference on Poverty Alleviation held in the capital city of Yunnan Province, China, in July 2010. We reaffirm our will to enlarge the objective area of the proposed Fund into a Global Anti-Poverty Fund following consultations with our intra-regional political partners in Latin America and the Caribbean under COPPPAL and with the political parties in Africa.

We realize that, as Asia's economies expand, the continent's huge population will exert an upward pressure on food demand worldwide. To anticipate this, we must now build up our centers of sustainable agricultural growth. The Mekong River basin can become one such resource, both for Southeast Asia and beyond. We urge the Mekong Basin states to protect their water resources and develop their agriculture — particularly its organic components — not only to reduce poverty among their peoples, but also to improve regional food security and to restrain the inflation of global food prices.

Dealing With the Perils of Climate Change

We recognize our critical need to turn our economies away from business as usual — so that we can face up to our collective responsibility to deal with the hazards and perils of climate change. Already our countries must cope increasingly with ecological disasters such as super-typhoons, great floods and engulfing landslides. We must reexamine our accustomed energy culture and move forward in the innovative use of renewable and clean energy sources.

We must strengthen the bonds between humankind and the living earth by making industry a friend and not an enemy of the environment. Public policy must encourage the rise of 'green enterprises' and the adoption of technologies attuned to the new environmental imperatives. We must always keep in mind the earth is not for us to deal with as we please. We're merely its trustees for the next generation.

Toward Renewable Energy

We recognize that moving toward the use of renewable energy is not just an effort to reduce CO_2 emissions. It is also an opportunity to create sustainable domains of economic activity, raise popular incomes and generate jobs. The judicious development of Asia's renewable energy resources — in all their variety and widespread availability — should become the end-goals of our continent's energy strategy.

Since it will take some time for renewable energy resources and green technologies to become fully developed, we must — at least for the foreseeable future — continue to rely on conventional energy sources to fuel Asia's development. Hence we must also advocate — in an era of volatile oil prices — the tapping of the oil and gas reserves known to lie both inland and offshore in many regions of our Continent.

The Role of Biological Diversity

We recognize that biological diversity has played a crucial role in the rise of civilizations. Humankind must preserve zealously this immense biodiversity reserve. We call for the founding of an 'Asian Eco-Safety Research Center' to support our Continent's sustainable development and to encourage pioneering and innovation in sustainable development, particularly in farming, forestry and fishery.

Coping With Natural Disasters

We also call for the closest cooperation among the Asian states to deal with recurring natural disasters, such as earthquakes, tsunami and flooding. We must link all efforts to prevent these disasters and to provide relief and rehabilitation at all levels — local, national and international. We urge the convening of an ICAPP Medical Emergency Forum (IMEF) to respond to the challenges and threats from natural disasters, as agreed at the 1st Meeting of the IMEF Steering Committee in Langkawi, Malaysia in May 2010. We welcome the decision to convene the ICAPP Conference on Natural Disasters and Environmental Protection in Malaysia in May 2011. We approve the (Angkor) Protocol of World Eco-Safety Assembly, and support objectives of the International Eco-Safety Cooperative Organization and propose that we show our collective resolve and consider including in our respective national Constitutions and political party platforms provisions and solutions to combat climate change and environmental degradation.

Youths' and Women's Organizations

We recognize that youth and women's access to and control over resources, access to opportunity and participation in al sphere of life are vital to sustainable development and prosperity. We recognise that the social-cultural processes of modernisation cannot be completed without their active engagement.

Mobilize Political Parties for Negotiated Settlements

We seek reconciliation in all Asia's conflict zones. Indeed one of our end-goals is to mobilize all the world's political groupings on behalf of the global peace process; the political and economic integration of the regions; the stability of the economic order; debt relief, and common action against poverty, corruption and climate change. By getting together and agreeing on common action, we, the Asian and world's political parties, could contribute our share to bridging the East-West, North-South divide; prevent the threatening "clash of civilizations," and save our planet from environmental degradation.

Preserving the Asian Heritage

We must also emphasize the importance of preserving Asia's splendid heritage, its cultures and values to the extend that they do not violate the rights of any individual and group that even under globalization, continue to inspire and to guide our Continent's processes of modernization. We call for the creation of an institution to award grand prize for union and innovating contributions to sustainable development, particularly in the field of agriculture, environment, energy, peace and culture.

Felicitate the Royal Government of Cambodia and the International Coordinating Committee (ICC) for their achievement with respect to the Angkor World Heritage Site, a cooperation between 18 countries and 25 institutions in 60 different projects, a model of excellence in its execution of protection, preservation and sustainable development and harmonization between cultural heritage, nature and human activities.

Praise for the Cambodian Leaders

Under the auspices of His Majesty King Norodom Sihamoni and His Majesty King Father Samdech Preah Bat Norodom Sihanouk, we compliment the Cambodian People's Party for achieving the following goals:

— Liberating the country from the genocidal regime, preventing its return, and rebuilding the country from scratch with poverty rate of 100 percent;

— Negotiating and achieving a political settlement which paved the way for the development and prosperity;

— Safeguarding the Constitution and, thanks to win-win policy of Samdech Techo Hun Sen, guaranteeing peace, political stability and full national unity that has enabled the economy to growth by double digits in recent years;

— Brining the last Khmer Rouge leaders to justice through an Extraordinary Chamber within the Courts of Cambodia (ECCC) that has been recognized by the United Nations as a new model for the international justices; and

— Dispatching Cambodian troops on UN peace-keeping mission to a number of countries such as Sudan, Chad, Central Africa and Lebanon.

Within a short spam of less than two decades, Cambodia has demonstrated through its policy a model of peace and reconciliation. A large-hearted approach based on inclusion, generosity and taking all on board has ensured a process where there no loser and all are victors. Such a model has relevant for others conflicts in Asia. In this context, ICAPP supports and joins the Centrist Asia Pacific Democrats International (CAPDI) Peace Commission in its efforts to promote peace in Asia through dialogue among political parties in conflict areas such as Nepal, Korea, Pakistan-India, and Afghanistan.

The ICAPP will be sending a delegation to Kathmandu on the invitation of the three major political parties of Nepal to seek a way out of the impasse through dialogue among the parties concerned.

The ICAPP also welcomes the establishment of the High Peace Commission by the Government of Afghanistan, which is mandated to start negotiation with the insurgents with a view to seeking a political settlement of the conflict.

Pursuant to the letters written by the ICAPP Chairman, Jose de Venecia and Secretary General Chun Eui-yong, to the President of Pakistan and the leader of the ruling Indian National Congress, Madame Sonia Gandhi, representatives of the Indian and Pakistani ruling, and opposition parties met in Phnom Penh on December 3, 2010, and agreed to institutionalized the dialogue amongst the two countries main political forces.

With respect to the recent provocation and military action in the Korean Peninsula, the ICAPP General Assembly calls on all parties concerned to immediately defuse the situation through dialogue and negotiations. The ICAPP strongly urges the international community to ensure that there is no recurrence of the use of force, in line with the statement of the UN Secretary General on this issue.

The ICAPP General Assembly noted that the "Cambodian Model" of peace, reconciliation and integration of armed groups presents a way forward to resolve conflicts in other parts of Asia.

Acknowledgments

Lastly, we must express our gratitude to the Cambodian People's Party and the FUNCINPEC Party for hosting our Sixth General Assembly. Further, we thank the Royal Government and the people of Cambodia for their warm hospitality. And we note with gratitude the encouragement and support of the Hanns Seidel Stiftung and the Konrad Adenauer Stiftung, both of the Federal Republic of Germany; and of the Korea Foundation, Republic of Korea.

We also express our gratitude to the representatives of political parties and institutes from Argentina, Columbia, Germany, Mexico, Seychelles, Sudan, Sweden, Tanzania, Uganda, and the United States who have participated in the assembly as observers, and to the delegation of International Eco-Safety Cooperative Organization. The ICAPP General Assembly paid tribute to the creative leadership and dynamism of His Excellency Jose de Venecia as the architect and moving spirit of the ICAPP.

Appendix VI :

(Angkor) Protocol of World Eco-Safety Assembly

Phnom Penh, Cambodia December 3, 2010

Contracting parties of the Protocol,

Being delegations of political parties, parliaments and governments from Asia, South America, and Africa present at the 6th General Assembly of ICAPP and the 1st World Eco-Safety Assembly.

Having noticed that recent years have witnessed the world's worst ecological disasters one after another, such as earthquake, tsunami, hurricane, volcanic eruption, flood, drought, mud-slide, forest fires, land desertification, sandstorm, water resources contamination, environmental contamination, industrial contamination, sharp decline of wetlands, species extinction, destruction of marine ecosystem caused by crude oil spill, and epidemic diseases such as Aids, SARS, bird flu, mad cow disease, foot and mouth disease, cholera; and those ecological disasters caused by climate change and human economic activities have posed a serious threat to human survival and national security.

Having realized that Angkor civilization, together with Babylonian civilization, ancient Egyptian civilization, ancient Indian civilization, and ancient civilization of the Yellow River, is an important component of world civilization; recent years, political parties and governments of every country have made significant contribution to the fields such as the promotion of ecological civilization, coping with climatic change, maintenance of ecological safety, environmental protection and realization of

UN Millennium Development Goals,

Having further realized that maintenance of eco-safety, poverty elimination, efforts to cope with unexpected ecological disasters, attainment of coordinated development of economy, society and ecology need the close cooperation and coordinated and concerted action of political parties, parliaments, governments and all walks of life all over the world,

Appeal that:

Efforts should be made to deepen the research on the eco-safety theory, and help global citizens truly realize eco-safety (that is, assurance of survival) is dynamic process in which the environment (air, soil, forest, vegetation, ocean and water etc.), necessary for national survival and development, is free or less free from destruction; the contracting parties shall bring eco-safety into national educational system;

Step should be taken to stick to the principle of harmonious development between ecology and economy; mode of economic development should adapt to the ecological carrying capacity; permanent utilization of resources and virtuous cycle of ecosystem are the utmost important objectives of development; when legislating laws and related polices, care should be taken to proceed from each nation's actual conditions and to take both the urgent need of protecting ecological safety, practical and feasible economy and social impact into consideration; care should be taken to respect each country's condition and the sustainable development modes of their own choosing according to their own development conditions; care should be taken to promote the balanced development of ecological and economic globalization in the whole world;

It is imperative to share responsibilities of eco-safety on a fair and rational basis; presently, the world eco-safety situation, especially the ecological environment in the developing countries and the less developed countries, is continuously deteriorating; concerted efforts of all countries are crucial to the maintenance of eco-safety and realization of the UN Millennium Goals; the principle of "Common and Differential Liability" should be carried out, by which the developed countries should actively and positively bear the responsibility of the maintenance of eco-safety and poverty reduction, and play more significant role in those two aspects, whereas the developing countries and the less developed countries should combine ecological construction with economic development, set up eco-safety early warning mechanism, and jointly cope with unexpected ecological disasters;

It is imperative to remove the obstacle in the exchange of ecological technology; sustainable development relies on the development in the fields of technology, information, biology and energy, which is of significance in promoting the efficient use of resources and keeping eco-safety; for the developing countries and the less developed countries, introduction of technological and managerial experience counts for as much as the introduction of capital; governments of all countries and international organizations should play an important role in promoting technological cooperation, and should protect intellectual property rights as well as establishing sound technology-transfer mechanism, promote the international exchange and cooperation of eco-safety system and poverty reduction technologies;

It is imperative to create favorable environment for international cooperation; steps should be taken to build a good new international economic and trade order which calls for concerted efforts of every country; the developed countries should further open their markets and reduce or even abolish the trade barriers resulted from the over high standards of eco-safety so as to promote the synchronized pace of development between eco-safety and international trade; the developing countries and the less developed countries should continuously step up the effort in alleviating poverty, raise their capacity of sustainable development, and take an active part in the international cooperation and competition,

And also appeal that:

In the opinion of Contracting parties of the Protocol, national leadership and coordination institutions on eco-safety should be set up; ecological disasters may break out at anytime under the current circumstances of mass migration and rapid urbanization; therefore, ICAPP agrees, with IESCO and contracting states, to jointly set up International Emergency Rescue Organization, and establish Fund of Saving Earth Homeland by World, aiming to raise the governments, society and the public' awareness on disaster prevention and build up the self capability of emergency rescue and aid by various effective means;

The contracting parties of the Protocol agree that both World Eco-Safety Assembly and World Expo on Eco-Safety shall be held biennially; the parties also agree to set up "Special Contribution Award for Maintaining Eco-Safety", and establish International Eco-Safety Certification Management Center in Hong Kong according to international eco-safety management system to carry out the certification and management of eco-safety city, park, tourist area, community, technology and place of origin of product,

And hold that one important objective of international natural disaster prevention and reduction is to reduce the frequency of natural disasters that happen in the developing countries and the less developed countries; it is imperative for the contracting parties of the Protocol to carry out wide-ranging international cooperation and exchange, hold forum and publicity events in International Day for Natural Disaster Reduction, and spread the disaster prevention technologies that individual countries gained in natural disasters and unexpected ecological disasters to countries all over the world, such as early warning of natural disasters like hurricane, tsunami, flood, snow disaster, volcanic eruption, earthquake, landslide, mud-slide, and so on, as well as successful and unsuccessful experience in disaster reduction and prevention,

And hold that the exploitation of ecological system (forest ecosystem, wetland ecosystem, marine ecosystem, basins of international rivers) in cross-border areas should strictly abide by the international conventions such as United Nations Convention to Combat Desertification, United Nations Convention on Biological Diversity, Convention on Wetlands of International Importance, the Helsinki Rules on the Uses of the Waters of International Rivers to ensure rational utilization and scientific exploitation; every country's legislative body should enact laws on ecological safety and environment; the institutions and responsible persons should be held accountable for the economic and legal responsibilities due to their acts of destroying ecological environment and endangering ecological safety, which consequently bring about heavy casualties, significant property damage,

And also hold that if any dispute over the explanation and application of the Protocol or accessory documents of the Protocol arises, the dispute shall be settled through negotiation and peaceful dialogue.

IESCO is the executive body of (Angkor) Protocol of World Eco-Safety Assembly. This Protocol has passed the deliberation of the Sixth General Assembly of the ICAPP and the First World Eco-Safety Assembly.

Appendix VII :

Jiang Mingjun Was Elected as CEO of CAPDI Climate Change Committee

On December 1st in Cambodia, CAPDI (Centrist Asia Pacific Democrats International) issued Phnom Penh Peace Agreement, which called on the various countries in the world to ensure the regional peace, safety, stability and prosperity, to take the attitude of mutual nonaggression and noninterference in handling international affairs, to solve territorial disputes by peaceful means, and to observe international treaties and international laws.

The above-mentioned Peace Agreement was passed by CAPDI at its first executive Board of Directors meeting held on the morning of December 1st at the Conference Hall of Cambodian Prime Minister Office. Representatives of political parties for signing the agreement included Jusuf Kalla (former Chairman of the Golkar Party of Indonesia), Jose de Venecia (former Speaker of the House of Representatives of the Philippines), Mushahid Hussain Sayed (Secretary-General of the Pakistan Muslim League), Konstatin Dosachev (Chairman of the International Affairs Committee of the United Russia Party), Sok An (Deputy Prime Minister of Cambodia and member of Standing Committee of People's Party), Keo Puth Reasmey (Chairman of Cambodian Funcinpec Party), Jiang Mingjun (Director-General of IESCO), Dr. Chuai Soi Lek (President of Malaysian Chinese Association) and Chuang Eui-yong (former Chairman of Foreign Relations Committee of United Democratic Party of Republic of Korea), etc.

"We are for disarmament and the non-proliferation of weapons of mass destruction, and we are against all forms of extremism, separatism and terrorism."

The organization promised to maintain the maritime safety, to support the Declaration on the Conduct of Parties in the South China Sea, and to call on all the contracting parties to observe it sincerely.

In addition, focusing on United Nations Millennium Development Goals, Phnom Penh Peace Declaration will establish Asia Poverty-relief Fund and Asia Petty Loan Fund, render debt cancellation and debt relief to the most impoverished countries, and air views on environmental protection and new economic ideology, etc.

Through all-dimensional cooperation and dialogue among centrist political parties, such as ICAPP, APA and other regional political parties, the representativeness of Asian political parties is expected to be expanded.

Hun Sen: Creating a More Beautiful Asia

When attending the signing ceremony of Phnom Penh Peace Agreement, Cambodian Prime Minister Hun Sen expressed that the predecessor of CAPDI was CDI-AP. At the executive Board of Directors meeting held yesterday, this organization changed its name and passed new regulations in order to further deepen and expand its engagement with continuously changing political environment.

"The newly passed regulations will encourage more democratic and civil social organizations, NGOs, think tanks, scholars, enterprisers, media, women and youth groups to participate in it to promote our influence in political, economic and social fields."

"I believe that the peaceful coexistence of political parties and countries with different ideologies could create a more beautiful Asia. In the era of high globalization, it is necessary to promote mutual trust of different political parties."

He was also grateful to CAPDI for appointing him as the honorary chairman and has committed himself to giving help and support to the best of his ability.

Looking For Peaceful Solutions by Means of Dialogue between Political Parties

Jose de Venecia, former speaker of the House of Representatives of the Philippines, expressed that CAPDI would focus on political contradictions and conflicts in Asia, and try to look for peaceful solutions by means of dialogue between political parties.

When interviewed by this magazine, Jose de Venecia said, Mushahid Hussain

Sayed, Secretary-General of the Pakistan Muslim League-Q, Pakistan, had been appointed as the Secretary-General of this organization at present and he would pay attention to Afghanistan conflict.

Jose de Venecia, founding chairman of International Conference of Asian Political Parties (ICAPP), pointed out that CAPDI would cooperate with ICAPP to mediate for the current political deadlock in Nepal and conflict between DPRK and South Korea.

Besides, Doctor Chuai Soi Lek who is Chairman of Malaysian Chinese Association said, as the executive director units of CAPDI were political parties, this organization can go beyond the national boundary and help the political parties and countries with different ideologies to carry on dialogues and communication.

"We strongly oppose all forms of racial extremism and this is very important to guarantee the peace and stability in Asian countries with racial diversity."

This assembly unanimously elected Dr. Jiang Mingjun, Director-General of International Eco-Safety Cooperative Organization, as the CEO of Climate Change Committee of ICAPP.

Appendix VIII :

IESCO Becomes the International Observer of the ICAPP

On the 6th General Assembly of the ICAPP which just concluded, IESCO was established as the international observer of the ICAPP.

In recent years, IESCO dedicates itself to the cooperation with political parties, parliaments and governments of various countries, and has made significant contributions to promoting peaceful development, maintaining ecological safety and realizing UN Millennium Development Goals. In 2009, IESCO attended the 5th General Assembly of the ICAPP in Kazakhstan, participated in the drafting of Astana Declaration, and provided the Asian political parties with strategic consultation in climate change, ecological safety and environmental protection. In this year, IESCO also participated in the Standing Committee Meeting of the ICAPP held in Nepal, and took part in the ICAPP Conference on Poverty Alleviation in Yunnan. At present, IESCO has become an important strategic partner of the ICAPP, COPPPAL and CAPDI. On Dec. 1, 2010, at the Centrist Asia Pacific Democrats International Conference, Jiang Mingjun, Director-General of IESCO, was elected the Council Member of CAPDI, and was made the Chief Executive Officer of CAPDI Climate Change Committee. Hun Sen, Prime Minister of the Kingdom of Cambodia, and Jose de Venecia, Former Speaker of the House of the Representatives of Philippines and Founding Chairman of ICAPP, were appointed as the Honorary Chairman and Chairman of CAPDI respectively.

On Dec. 5, 2010, as observers of the ICAPP, Jiang Mingjun, Director-General of IESCO, and Tee Ching Seng, Secretary-General of IESCO, together with representatives of the ICAPP Standing Committee, went to Preah Vihear from Siem Reap by helicopter. They inspected the areas of conflict along the border of Thailand and Cambodia, witnessed the signing of Phnom Penh Peace Agreement, which realized the wish of regional stability and peaceful development.

Appendix IX :

Jiang Mingjun and Tee Ching Seng Were Granted the "ICAPP 10th Anniversary and Service to Humanity Award"

During the 6th General Assembly of the ICAPP and the 1st World Eco-Safety Assembly held in Cambodia, on December 3, 2010, H.E. Samdech Hun Sen, Prime Minister of the Kingdom of Cambodia, Hon. Jose de Venecia, Jr., Co-Chairman of the ICAPP and Hon. Chung Eui-yong, jointly issued an order confering the "ICAPP 10th Anniversary and Service to Humanity Award" on Dr. Jiang Mingjun, Director-General of IESCO and Dr. Tee Ching Seng, Secretary-General of IESCO, in recognition of their outstanding contribution to the promotion of peaceful development and the maintenance of ecological safety.

Political parties, parliaments and government delegations from various countries of the world attending the 6th General Assembly of the ICAPP and the 1st World Eco-Safety Assembly took part in the Award-Giving Ceremony.

Appendix X :

Special Contribution Award for Maintaining Eco-Safety

According to the *Phnom Penh Declaration* of the 6th General Assembly of ICAPP and *(Angkor) Protocol of World Eco-Safety Assembly*, IESCO has established "Special Contribution Award for Maintaining Eco-Safety" in order to commend and reward those national governments, nongovernmental organizations, academic institutions and government leaders who has made important contributions towards maintaining eco-safety, protecting ecological environment, dealing with climate change and realizing United Nations Millennium Development Goals.

This biennial award will hold its award ceremony at every World Eco-Safety Assembly. The first award-winning institutions and government leaders are:

1. H.E. James Michael, President of the Republic of Seychelles;

2. Dr. Anna Tibaijuka, UN Under-Secretary-General and Minister of Ministry of Lands, Housing and Human Settlements Development;

3. Government of the Republic of Maldives;

4. The Government of the Kingdom of Cambodia;

5. International Academy of Ecology and Life Protection Sciences (IAELPS).

During the 1st World Eco-Safety Assembly, both Kansas City of the United States and Richmond City of Canada were awarded the honorable title of "International Eco-Safety Demonstration City"; and Tangxia Town, Dongguan City, China was awarded the honorable title of "International Eco-Safety Demonstration Town". Jiang Mingjun, Director-General of IESCO, was granted the medal of honor "Scholar Stars" by IAELPS. And Gennady Shlapunov, Deputy Director-General of IESCO, won

Chinghiz Aitmatov Golden Medal.

H.E. Mendsaikhan Enkhsaikhan, Former Prime Minister of Mongolia, and Co-chairman of IESCO, Dr. Anna Tibaijuka, UN Under-Secretary-General, Minister of Land, Housing and Human Settlements Development of Tanzania, and Co-chairman of IESCO, and H.E. Zhang Deguang, First Secretary-General of the Shanghai Cooperative Organization and Co-Chairman of IESCO, conferred award certificates and medals on the above-mentioned award-winning institutions and government leaders.

Resumes of the winners

James Michel: He is now President of The Republic of Seychelles. Since H.E. James Michel's taking office, Seychelles has signed various international treaties related with environment; issued a constitution to protect the right of every resident to enjoy and live in a clear, healthy and ecological balance environment; imbued every citizen of Seychelles with the concept that the harmonious co-existence of human and nature is everyone's responsibility in order to produce in them a natural love of nature and the self-control in the process of development. Under these measures, in Seychelles, the natural environment has received less damage in comparison with other places, and human has kept a close, mild and humble relation with nature.

Anna Tibaijuka: She was born in Tanzania and is a PHD of agricultural economics. In 2000 she was appointed as the Executive Director of UN-HABITAT by UN Secretary-general Ban Ki-moon. In 2002, UN General Assembly upgraded UNCHS (United Nations Centre for Human Settlements) and then established the present UN-HABITAT. During her term of UN Under-Secretary-general and Executive Director of UN-HABITAT, she strengthened the functions of UN-HABITAT from political, financial, and operation aspects, expanded the influence of UN-HABITAT, made UN-HABITAT play a more important role in international development and pushed forward many developing countries to take on the path of poverty elimination and sustainable development. On September 2010, she was relieved of her office of Executive Director of UN-HABITAT.

Government of the Republic of Maldives: The Maldives located in the India Ocean are regarded by travelers as the "Heaven" and are considered to be the happiest place under the sun. The prediction that the Maldives may vanish due to climate warming has made it keep on its toes in facing the ecological environment. In an effort to protect the environment, the Maldives established its first National Action Scheme for Environmental Protection which laid the foundation for preventing ecological crises in the future and restricted the excessive exploitation of natural resources. In order to attract the attention of the international society, the Maldives held the first underwater cabinet council in the world. The president and the 11 cabinet officials dived into the 6 meters deep underground and, by this intuitive action, called for countries across the world to take actions to deal with climate warming.

The Government of Cambodia: Under the leadership of the Cambodian government, the whole country has made a positive response to the climate change, improved conditions of the use of resources, implemented bio-diversity conservation, popularized the education of environmental protection, started environmental evaluation and developed eco-tourism projects. In order to protect its forests, it drew up the Forest Law regulating that the forest exploitation must be in line with a planned and sustainable way. In order to protect the water resource, in recent years, the Cambodian government has taken various measures to enhance its management, exploitation and application of water resources, making sure of the balance of water-using at present and in the future and the sustainable development of economy and society.

International Academy of Ecology and Life Protection Sciences (IAELPS): The IAELPS was established by Commonwealth of Independent States (CIS) and International Federation of Academy of Sciences in St. Petrsburg according to the resolution of United Nations Conference on Environment and Development held in Rio de Janeiro of Brazil in 1992. During the 18 years after its establishment, the IAELPS has committed itself to uniting those excellent scientists, scholars and engineers engaging in the study of ecology and life sciences of countries all over the world, and made remarkable contributions in advancing technological development, carrying out more important and promising scientific study, preserving ecological safety and life security and safeguarding the health of humanity. At the present time, the IAELPS has established cooperative relations with 64 countries and set up representative office and men by permanent delegate at the General Headquarters of the United Nations in New York.

Kansas City: The Kansas City, which is located in the Central America, has always strived to reduce the use of fossil fuels by using its rich local wind energy resources and developing economy depending on the renewable energies, in an effort to enhance its energy independency, protect environment and provide its citizens with a healthy living space. As an agricultural city, Kansas City has actively implemented the subsidy program of agro-ecology protection and strengthened the protection of agro-ecology in order to preserve the agro-ecology environment. It has helped the farmers carry out measures of long-term vegetation protection including the return of farmland to forest and grassland etc., with the view of ultimately reaching the goal of improving water quality, controlling soil erosion and bettering the ecological environment in which wild plants and animals live.

City of Richmond: Richmond is situated in the west coast of Canada and at the mouth of Fraser River. With a principle of priority-of-protection and the advanced management philosophy, it has worked out the urban planning by using ecology principle, and attached great importance on the education of "environmental consciousness" and "environmental culture". With the purpose of furthering the protection of ecological environment, Richmond has developed vigorously the environment-friendly industries, enhanced the comprehensive application of resources and paid great attention to the development of green industries and environment-friendly technologies. In the aspect of energy using, it encourages the using of clean energy such as wind energy, solar energy, and tide energy etc, as well as encourages its city residents to choose public transportation or bikes. Owing to these efforts, the City of Richmond becomes the example of ecological safety city.

Government of Tangxia Town: The Tangxia Town is situated in the Pearl River Delta of South China. During the process of industrialization and urbanization, it has always stuck to the development concept which lays equal emphasis on economy and environment and promotes harmonious coexistence between mankind and nature. While pushing forward the economic development with all its strength, Tangxia Town attaches great importance to the protection of natural ecology and development of social and public services, makes great efforts to carry out "green GDP project, blue sky project, clean water project, friendly living environment project and green land project". In this way, Tangxia Town constructs a dynamic eco-safety system with surrounding ecological green land, circulation of green economy and harmonious civilization. Tangxia Town has been awarded the following titles: China National Garden Town, China Famous Green Town, China Beautiful Town and China Hygiene Town, and so on.

Jiang Mingjun: He was born in Qingdao City, Shandong Province, China, who is the Vice President of UN/International Academy and Ecology and Life Protection Sciences, Chairman of UN Initiative and Technology for the Youth and Director-general of International Eco-Safety Cooperative Organization, and a PHD majoring in ecological safety. In 1998 he put forward the new strategy of "bringing eco-safety into national development." In 2006, he initiated the establishment of International Eco-Safety Cooperative Organization and regarded eco-safety maintenance, ecological environment protection, ecological conservation promotion and realization of sustainable development strategy as his duty. His major research achievements include Ecological Safety: Foundation of National Survival and Development and Comprehensive Introduction on Eco-Safety, which have laid academic foundation for ecological safety theory.

Gennady Shlapunov: He is now the Assistant Director-General of International Eco-Safety Cooperative Organization (IESCO), Editor-in-Chief of the International Ecology and Safety magazine, Vice President of the Writers League of Eurasia and Member of the Writers Union of Russia. After working hours, he wrote a collection of poetry Ecology of Feelings in 50 years time. This is a remarkable poetical works and a long-lasting song of praise dedicated to all lovers under the sun.

Appendix XI : *

Jiang Mingjun and His Monograph
Comprehensive Introduction on Eco-Safety

Mr. Jiang Mingjun was born in Qingdao, Shandong Province. He is the Vice President of UN/International Academy of Ecology and Life Protection Sciences, Vice-chairman of World Eco-Safety Assembly, Director-general of International Eco-Safety Cooperative Organization and Chaiman of United Nations Initiative and Technology for the Youth.

The Monograph on Comprehensive Introduction on Eco-Safety embodies Mr. Jiang Mingjun's 14 years painstaking work. In the foreword, he overturns the traditional concept and first of all, redefines eco-safety as "foundation of human survival and national development which is not merely about the environmental protection, but an equivalent to political security, military security and economic security".

The book is divided into two parts: basic research on eco-safety and eco-safety

---

* Pictures in this article are covers of Chinese and English edition of *Comprehensive Introduction on Eco-Safety*.

management system. The first part expounds the concept and connotation of eco-safety, with an emphasis on the analysis of the national eco-safety. The second part moves from theory to practice. It introduces particularly the international eco-safety management system. This will give much guidance and reference to the construction, management, evaluation and assessment of eco-safety city (town), eco-safety community, eco-safety base and eco-safety product.

Appendix XII : *

Gennady Shlapunov and His Poetry Anthology *Ecology of Feelings*

Gennady Semenovich Shlapunov is from Russia and present Assistant Director-General of International Eco-Safety Cooperative Organization, the Coordinator of Russian National Committee for UNEP, Vice President of the Writers League of Eurasia, Member of the Writers Union of Russia, Editor-in-Chief of the International Ecology and Safety magazine, academician of United Nations International Academy of Ecology and Life Protection Sciences. Besides, Chinghiz Aitmatov Golden Medal was given to Mr. Shlapunov by the Golden Medal Global Committee.

The poetry anthology Ecology of Feelings is the crystallization of the poet's creative essence for over 50 years. In this anthology, the poet extends and sublimates his comprehension of love and life to the concerns of mankind and ecological emotions of the earth. Protecting lives on earth is the gist of feelings, soul and hope of ecology. The money earned through the issue of this poetry anthology will be used for the environmental protection undertakings of the United States and charitable activities.

The poet's third poetry anthology Moments and Eternity is about to be published.

---

*　Picture in this article is cover of *Ecology of Feelings*.